# EMBODIED TECHNICS

Other books from Automatic Press ♦ $\frac{V}{I}P$

**Formal Philosophy**
edited by Vincent F. Hendricks & John Symons
November 2005

**Masses of Formal Philosophy**
edited by Vincent F. Hendricks & John Symons
October 2006

**Political Questions: 5 Questions for Political Philosophers**
edited by Morten Ebbe Juul Nielsen
December 2006

**Philosophy of Technology: 5 Questions**
edited by Jan-Kyrre Berg Olsen & Evan Selinger
February 2007

**Game Theory: 5 Questions**
edited by Vincent F. Hendricks & Pelle Guldborg Hansen
April 2007

**Legal Philosophy: 5 Questions**
edited by Morten Ebbe Juul Nielsen
October 2007

**Philosophy of Mathematics: 5 Questions**
edited by Vincent F. Hendricks & Hannes Leitgeb
January 2008

**Philosophy of Computing and Information: 5 Questions**
edited by Luciano Floridi
Sepetmber 2008

**Philosophy of the Social Sciences: 5 Questions**
edited by Diego Ríos & Christoph Schmidt-Petri
September 2008

**Epistemology: 5 Questions**
edited by Vincent F. Hendricks & Duncan Pritchard
September 2008

**Mind and Consciousness: 5 Questions**
edited by Patrick Grim
January 2009

See all published and forthcoming books at
www.vince-inc.com/automatic.html

# EMBODIED TECHNICS

Don Ihde

Automatic Press ◆ $\frac{V}{I}$P

Automatic Press ♦ $\frac{V}{I}$P

Information on this title: www.vince-inc.com/automatic.html

© Automatic Press / VIP 2010

This publication is in copyright. Subject to statuary exception
and to the provisions of relevant collective licensing agreements,
no reproduction of any part may take place without
the written permission of the publisher.

First published 2010

Printed in the United States of America
and the United Kingdom

ISBN-10   87-92130-27-5   paperback
ISBN-13   978-87-92130-27-3   paperback

The publisher has no responsibilities for
the persistence or accuracy of URLs for external or
third party Internet Web sites referred to in this publication
and does not guarantee that any content on such
Web sites is, or will remain, accurate or appropriate.

Typeset in $\LaTeX 2_\varepsilon$
Photo and graphic design by Vincent F. Hendricks

# Contents

1 Introduction: The Question Concerning Embodiment    iii

2 Technofantasies and Embodiment    1

3 Technologies—Musics—Embodiments    17

4 Phenomenologists and Robots    37

5 Beyond Embodiment and a Return    55

6 IT: Clouds and Cyberspace-Time    69

# 1

# Introduction: The Question Concerning Embodiment

In *Embodied Technics* I shall be addressing a number of perspectives on our embodied and mediated experience with and through contemporary technologies. Today, in media studies, science studies, cultural studies—and in much philosophy—there is deep attention to questions concerning the human-technology interfaces we all experience. And not all thinkers agree: some hold that the new technologies of media, imaging, and digital-computational technologies *disembody* the human. Such technologies are thought to take us away from ordinary and face-to-face experience and *distance* us from others, nature or even objects. Others hold that our bodies are actually reducible to machinic bodies, such that our 'minds' can be downloaded onto computers or be 'brains-in-vats.' They argue that it is only a matter of time until this will happen. And others, often coming from phenomenological or postphenomenological positions, argue that our contemporary technologies actually *embody* or *re-embody* our fleshly experience in new ways, in interactive ways.

This book, while sometimes discussing the arguments concerning embodiment, clearly sides with those who examine the human experience of embodied technics and try to discover the new ways in which contemporary experience is transformed through new or re-embodiments. There are many books which look at different aspects of embodiment, including Donna Haraway's concerns with situated knowledges in her *Simians, Cyborgs and Women* (1991), Kathrine Hayles, *How We Became Posthuman* (1999), Vivian Sobchack's *Carnal Thoughts* (2004) among them. But more recently I have been particularly inspired by Susan Kozel's dramatic and phenomenological work, *Closer* (MIT, 2007), which is a Merleau-Pontean exploration of her work as philosopher-dancer. She uses Merleau-Ponty's vocabulary, particularly that of his late work, *The Visible and Invisible*, to describe the new forms of embodiment in her performances. Analysing telepresence, she calls her account "a materialist account of space and virtuality." (1) I, too, have long called myself a "phenomenological materialist" by reference to the phenomena of embodiment. But this is not a reductive materialism or a mechanized materialism—it is rather a phenomenological and multidimensioned sense of body. And it is also an analysis of contemporary forms of embodiment through various media, imaging and digital computer processes. (2) Clearly my experience draws from different uses of these technologies than Kozel's, mine is more academic and far less bodily active, but the very liveliness of performance with its full spectrum of perception, motility, emotions, takes one to a horizon which can also be appreciated in cinema, music, science imaging and the other technological mediations I analyze here.

I begin with "Technofantasies and Embodiment," in the second chapter, which takes up the "Matrix Trilogy" in cinema. This set of movies, often become a favorite book and discussion topic among philosophers, draws from the most ancient of 'reality/illusion' and copy-theory traditions found especially in Plato, and then reformulated into image and representation 'theatre of the mind' epistemology in Descartes' tradition. I try to show how technofantasies are involved in these traditions and contrast these with a more phenomenological sense of embodiment. In the third chapter, "Musics-Technologies-Embodiments," while the focus is upon contemporary recording musics, I take a longer and more historical set of music technologies (instruments) and show the roles of changed embodiments in music histories. In chapter four, "Phenomenologists and Robots," I look at the roles of motility and perception in contemporary robotics with special emphasis upon the work of Hubert Dreyfus on artificial intelligence compared to the problems of human motility and sensory capacity. In chapter five, "Beyond Embodiment and a Return," I examine contemporary imaging technologies, particularly those developed in the sciences. These technologies, primarily invented only in the $20^{th}$-$21^{st}$ centuries now image along the entire electromagnetic spectrum from radiations which we embodied humans do not and cannot directly perceive. Yet, this new beyond-the-sensory imaging *is* experienced in a mediated way which leads us to new "worlds" never before perceived. And, finally, in chapter six, "IT: Clouds and Cyberspace-Time," I take the longest view of writing technologies. For although information and communication technologies are focal, I look at many more ancient practices which call for variant embodiments which precede our now global and networked set of expressive technologies.

The original presentations of each of these chapters is widely

1. Introduction: The Question Concerning Embodiment    v

divergent. Only the first two chapters have been previously published with the others new to this book. "Technofantasies and Embodiment," originally appeared in *The Matrix in Theory*, edited by Myriam Diocaretz and Stefan Herbrechter and published by Rodopi, 2006, reprinted here with permission by Rodopi. "Technologies-Musics-Embodiments" was originally published in Janus Head, 10.1, Summer/Fall, 2007, and is here reprinted with the permission of Janus Head. "Phenomenologists and Robots: Human and Machinic Embodiments," was originally presented in September, 2008, at Sendai University in Japan while on a research consultantship on a grant involving phenomenologists and roboticists. "Beyond Embodiment and a Return," was given at the Carlesburg Academy in Copenhagen, Denmark, in May, 2009, for a conference on "Shaping Knowledge," And "IT: Clouds and Cyberspace-Time" was given at a conference at the Free University of Brussels on "Autonomic Computing," January, 2009. I also wish to acknowledge the continued stimulation coming from the Technoscience Research Group at Stony Brook University, whose scholars produce studies of contemporary technologies and the impact therefrom on human experience. The next generation of philosophers, including Peter Paul Verbeek, Evan Selinger, Cathrine Hasse, Robert Rosenberger and others who are all already underway with studies of embodiment in different settings. And, finally, I am grateful to Vincent F. Hendricks and Automatic Press, the fastest and most radically contemporary style of publication in my experience! And, finally, thanks to Linda and Alissa for helping me tweak the computer.

**Endnotes:**

1. Susan Kozel, *Closer: Performance, Technologies, Phenomenology* (The MIT Press, 2007), p. xvi. [Susan Kozel participated in my "Imaging Technologies" Ph.D. seminar in Aarhus, Denmark where she presented several of her videotaped performances. ]

2. *Embodied Technics* follows my previous *Ironic Technics* with Automatic Press (2008). Prior to both, my *Bodies in Technology* (University of Minnesota Press, 2002) also traces an emphasis on embodiment through the disciplines of philosophy of technology, philosophy of science and science studies.

# 2

# Technofantasies and Embodiment

**Abstract:** Technofantasies and embodiment takes up the theme of how movies like the *Matrix* Trilogy play upon fantasy in a technological context and relate to the human sense of embodiment. The *Matrix* trilogy is related to its predecessors such as Plato's cave and the theatre of the mind, or *camera obscura*, which played the role of models for mind. Contemporary technologies are then used to explain some of the effects and implications for "mind" and embodiment in the *Matrix*. The conclusion is meant to suggest a different model of both technologies and "mind" than is usually taken for granted.

Fictions, whether written, staged, imagined, or made into movies, always call for a "suspension of disbelief." But there are degrees and styles of disbelief and I want to address some of these with respect to the *Matrix* trilogy. It all started with Plato and his wonderful anticipations of the movies in his allegory of the Cave. Of course, by today's standards, he was plagued by not having any hi-tech or anything even close to special effects. His "movie" was very lo-tech. His audience, the prisoners chained by the necks and immobilized on their benches, faced the cave wall, the *tabula rasa* or screen where the images would dance. Behind them was a fire – no projector or lens mediated system – in front of which was a parapet along which the hidden operators would hold up copies of copies, that is a sort of shadow puppet of ducks and rabbits or whatever, whose shadows would then be cast upon the screen-wall. But, it was audio-video in a primitive sense, because he described how an echo system made the noises seem to come from the cast shadows. Now, according to this artifice, and because the prisoners had been there since birth, his audience was supposed to "believe" that *reality itself* consisted of the shadow-play upon the wall. Here – Plato wants us to believe – is no suspension of belief; this *appearance-is-reality* in the belief of the prisoners. Plato, the inventor of the artifice, of course, knows better; he sees through the illusion and knows there is a very different hierarchy of realities and will show us how to find these by the gradual journey out of the cave into the outer, sun-lit world. But Plato's thesis is an ancient analogue to that of the "Matrix" – could an actual, human, embodied being be fooled into thinking that the cave's illusions, in the Matrix "program," can produce "real experience"? Or, indeed take programmed experience for lifeworld experience?

The role I want to play is that of *phenomenological skeptic* because I doubt that without a totally successful self-delusion, a total suspension of such belief is not possible. To try to demonstrate this, I will undertake several variations, first on the cave theatre, then on the later theatre of the mind which Descartes uses to update Plato, and on to the *Matrix* trilogy. To "believe" Plato – or Descartes – or the *Matrix* – what you have to suspend is your own *embodiment*. So, we return to Plato's cave: his theatre, as noted, is an audio-visual one, shadows on the wall, echoes in the chamber, a sort of reduced or minimalist display compared to what the prisoner would or could see outside the cave. Of course this sort of simulation and modeling has always been part of any abstractive strategy which uses the simpler to try to illuminate the complex. But, with respect to embodiment, the prisoners themselves are not simply audio-video beings – they are fully "bodied" and the implicit recognition of this is indicated in the need for their orientation to be fixed, immobilized by chains. They *cannot* turn around, they are tactile and kinesthetically immobilized as if they were only forward facing "eyes and ears." Of course, if they could turn around, they would immediately *see* the fire, the shadow-casting shadow puppets of ducks and rabbits, and realize, *perceive* in a *gestalt,* the causal situation of what produces what and thus confirms Plato's implicit metaphysics of whatever is more original is the cause of that which is dependent.

I would argue, that so long as the prisoners are fully embodied, that is they have the full range of sensory dimensions, and are aware of these as we all are, they simply cannot be fully fooled. Were I, as a prisoner, to try to turn my head, would I not realize I was being restrained? Would I not realize that, for example by turning my eyes from side to side, even if my neck is chained, there is a multiplicity of a limited sort of perspectives? Would I not be suspicious regarding the constraint system itself? My point is that so long as we have full, multidimensional embodiment, the awareness of constraint itself defeats any full illusion.

There is trickery here, a trickery like sleight of hand where the mechanism which produces the trick tries to remain itself hidden. Magicians – including Plato – of course themselves know the

## 2. Technofantasies and Embodiment

mechanism; they themselves do not believe in magic. But then, unless one is very young, neither do most audiences! Could we do better than Plato? Technologically – and I will claim psychologically – yes. For example we could do a hi-tech variation on the cave: a 3-dIMAX. Here, instead of ambiguous shadows on the wall, produced by an unfocused fire, we have full 3-d "images" moving around in front of us, maybe a shark with open mouth coming directly at us, or a flock of birds flying right by. Much more vivid, and surrounded by Dolby sounds. And, psychologically, Plato's captivity trick is not needed – we eagerly enter the theatre, willing to take our forward-oriented seats and can even move our heads since the surround-screen is so large its "illusion" is not hurt much by some head movement. But this is an improvement only by degree, since I was already outside and come in to be entertained and thus I already know something of the difference between the staged nature of the new theatre-cave.

And while this is a great leap in "realism" if you will, it is still not enough. Here the artifice of goggles reminds us of artifice, and even while watching and listening, if I stick my hand out to catch the "birds" they have no substantiality, and if I take off my goggles, the show turns to double-vision fuzziness. It is too easy to dispel the illusions. Both my sense of full embodiment, added to which I have the variation of outside of, versus inside of the theatre, keeps me from full suspension of disbelief. But, now we must turn backward not to antiquity, but to early modernity and a second variation upon a theatre of the *mind*.

The two best known early modern philosophers, John Locke and René Descartes, both used the same metaphorical device to describe a theatre of the mind – the *camera obscura*. This device, re-invented and used in the Renaissance mostly by artists, was known earlier by the Arabic philosopher, Al Hazen (1038) and was described in his *Optics*. And, the *camera obscura* can be said to simply be an optical version of Plato's cave with a few modifications. Dark room, *camera obscura,* has a small opening, by Locke's and Descartes' time including a lens, through which light enters. What is *outside*, whether a light itself, say the sun, or lighted objects, then casts an *image* on the opposite wall, Locke's *tabula rasa,*– for movies a screen, for Plato the cave wall – *inside*. To see this image one either has to *be inside* looking at the image, or have a second opening to look onto the screen from the outside. The inside image, however, is "reduced" from three-dimensionality on the outside to two-dimensionality inside, and is, moreover, inverted. Thus, like Plato's shadows, copies of copies, the *camera image* "represents" or copies the outside material [lighted] object. This optically produced image or illusion or appearance is a bit of an improvement on Plato's shadows since it is colored, "isomorphic" or is clearer and more spatially correct, but still a mere image.

The most dramatic change, however, introduced by both Locke and Descartes, is to have re-interpreted the prisoner, the observer of the images. Both took the *camera obscura* as a model for the *mind itself*. The mind, the seat of awareness, is a sort of observer literally inside the camera, who looks at the images and this is what becomes "subjective experience," the experience of experiencing one's own *thoughts*, mental images and the like. "Objects," in turn, are what is *outside* the camera, analogues to Plato's outside-the-cave, "real" objects. Now, here comes the rub: the newly enclosed "subject" *cannot get outside the box, but is essentially always inside it.* Now it is this move which much more deeply makes the plot illusion of the Matrix what it pretends to be.

For early modernity and the optical version of appearance/reality, the *theatre of the mind* is itself the real, the epistemology and metaphysics of this model. Note what is implied: first, early modernity introduces embodiment to a *dualism* of body and mind. The body in early modernity is "mechanical" and itself lifeless; the mind becomes "subject" and is a sort of homunculus inside the [body] box. In parallel fashion, "external" reality, the objects outside the box, are material, but the images or representations inside the box, are merely phenomenal, i.e., fleeting images and representations of the outside. Were this sustainable as the true description of reality itself, then the late modern leap to the Matrix version of the theatre of the mind would be simple. Instead of external objects casting the images upon the screen of the mind, it is a program which does so, then clearly the homunculus subject *could take the program as experience.* But the *Matrix* trilogy has its characters slipping into and out of the program – the threat of course is for the program to eventually become *total,* the machine wins, humanity defeated, another version of technology-as-Frankenstein.

With the Locke-Descartes theatre of the mind, does Plato get trumped? And how can I use my escape tactic by appealing to full embodiment here? The now separated body and mind pretends that even my tactility and kinesthesia are "sensations" merely caused by something external – my experience here is trapped "inside" and is "subjective." The device tries to persuade me that "I" am actually separated from my "body." My experience is not embodied.

It might seem, then, that I need a different tactic, so I shall try one. First, do the usual philosophical self-reflexive move: can Locke and Descartes be self-consistent with respect to their description? My answer is, "no!" Were *they themselves* to have consistently taken their version of the theatre of the mind *as reality*, and they themselves in the position described inside the box, *how could they ever think there is anything like "external" or outside reality?* Put another way, were they to have constructed this view of knowledge within the limits of their own description, could they have developed any distinctions of appearance/reality or image/object? Again, "no." Of the two, Descartes was the most sensitive to this dilemma. And to make the story as short as possible, note that his answer was posed in terms of a sort of philosopher's "God." For the poor enclosed "subject" to know truly would require knowing that there is a correspondence between the theatre images – all he has – and "real" external reality, that which is outside." Without going into the complex set of arguments he developed, what emerges as the guarantee of correspondence is the ideal viewing of "God" who can *simultaneously see inside and outside the box and thus judge the correspondence between entity and representation.* I call this a "cheat code." It is not "God" who sees both inside and outside the box, but Descartes, because he is the one who has described, invented, the theatre of the mind himself. He is outside the camera and can at the same time see inside it and that is what constitutes the full metaphysical and epistemic situation. But, one does have to admit that the trick is almost good enough to prevent my first tactic of trying to turn around in the cave and thus revealing my captivity through the experience of full embodiment. For while there is a "copy" of an external object in the *camera*, the image, the homunculus observer inside is not fully a body, were the observer to be so, he/she could presumably turn around and look out the peephole instead of at the *tabula rasa* and see for themselves what was out there. Descartes has distracted us but, exactly like Plato, Descartes is the inventor of his own artifice.

But, the cheat code position *is itself a position*, a situated position. And recognizing that, I contend we are back *in the world* (since we never left it) with our fully embodied positionality and actional, full body movement. In Merleau-Ponty's phenomenological sense, I find myself already outside myself *in the world*. Now we are closer to our entry into the Matrix world as well. As one final step before going to the movies, I will take a contemporary detour into a set of claims which might be equivalent to a "true believer" in a Matrix world. Hans Moravec has made something of an infamous reputation for himself and his fantasized robotry, by claiming that humans will someday (or even soon) download their minds into a computer. My question is: what kind of mind could be so downloaded? And my variant revolves around the phenomenological sense of embodiment: will I be able to download my body into a computer? And, in Moravec's sense, which would be harder to do? My guess is that by today's lights, downloading a human body into a computer might be thought to be more difficult than downloading a mind? But if this is a good guess, then it implies more about the late modern and still very "Cartesian" interpretation of mind than not. In early modern versions of dualism, mind is *immaterial* as mental substance, in contrast to a material body. In late modern variations the question of immateriality has become ambiguous, but the escape clause is one which interprets the mind – still separable from body – as more like an information processor. Which, if it were, would make it a likely candidate for "downloading." But a body, even in late modern parlance, remains in some sense, material, and so it is harder to conceive of "downloading" something material.

Embodiment, particularly in its phenomenological sense, is clearly material, or one might say experiencing material. Again, Merleau-Ponty's sense is that "I am my body." And there is no sense of some immaterial "mind." This would make downloading very difficult indeed. And, embodiment is also situated and relational; it entails action, interaction. So, if I "am" my body and am, interactionally "already outside myself" in the world then it equally becomes difficult to conceive of any total identity of image/reality. Here a nuanced difference between Plato and Descartes also emerges: Plato's prisoners can develop a doubt about the shadows by virtue of the difference between fully embodied self-awareness and the reduced presence of shadows; but the homunculus in the box presumably has no such experience because his/her body has itself been removed – but there is also then *no experience either of the outside*. Any difference is going to have to be strictly a matter of internal relations. It is only by virtue of the cheat code which must generate differences between inside and outside that any appearance/reality distinction could be made at all. And both Descartes and Locke employ this cheat code, albeit in slightly different ways. One way Locke does it is by making distinctions between perceptions and imagination. Imaginations are supposed to be more vague, less robust, less clear than perceptions since – again in parallel to Plato's shadow theatre – imaginations are supposed to be pale "copies" of perceptions. *And if the model is that of the camera obscura, that is true.* The image on the wall of the obscura is always less well lighted, less clear, and were one an artist doing a tracing inside one would find it hard to depict the full intensity of painting that which it images outside the camera. But only if one could compare these scenes, the internal and the external, could such a claim be made. The breaking of the early modern cheat code comes with the realization that I am already outside myself in the world. But even more radically, I would claim that imagination is *not* a copy of perception at all –while it can have some features of perception, it carries others which could be said to make perception a pale copy of the imagination! Even when imagining something as simple as a duck or rabbit, in imagination I can turn each into any color I want, including day-glo ones, make each grow or shrink, or even float them behind my head, characteristics which make perception seem quite restricted. Put another way, concrete imagining has different structural features than perception (cf. Ihde, 1973).

However, the other side of what I am taking as an inadequacy of Locke's theory of imagination, is the indirect evidence that he is following his own model of mind, the *camera obscura,* to the degree that the model perhaps overdetermines the description he gives of the phenomenon, that is, his descriptions sound more like what one would see in a *camera,* than of a description from someone with a vivid imagination.

We began with Plato's cave-theatre 2400 years ago; then jumped to the early modern theatre of the mind, now 400 years ago. As we now come into our own time take note that both these theatres were artifices, but also "technological set-ups." Now, my take on these theatres is that each in its own way served as an *epistemology engine,* that is, the inventors each used their technological models as a metaphor for how we gain knowledge and how we have experience. In our last historical leap, those same features endure, but in relation to a changed technological set-up, a new theatre. And, it is obvious that the cinematic hi-tech of the Matrix trilogy is dramatically different from the crude shadow puppetry of Plato

8    2. Technofantasies and Embodiment

and the less than clear and distinct upside down images of the *camera obscura*. In both the antique theatres, it was easy to see – assuming the position of one in "cheat code" position where one could compare the image world with the day light world – that these theatres were such that all images are both dependent upon the "real" things which are their cause and sources, and are dimmer and poorer than these sources. The fiction of our late modern theatre, hi-tech cinema, is quite different, its implicit claim is that its image world can substitute for or replace the ordinary lifeworld – at least while plugged in. The prisoner of the new cave equivalent or the homunculus in the camera subject now has all experience shaped by the new epistemology engine, the matrix program.

This new set of claims, actually remain homologous to those of Plato and Descartes but now differently technologically situated, fit into a changed metaphysical context as well. There are, with the Matrix, "two worlds" as in the predecessor versions, into and out of the program. The connector, that dramatic plug which is inserted into the installed receptacle in the back of the head-neck is the magic gateway between these worlds. And, the question of which is appearance, which reality gets played back and forth as well. If the characters are going to do battle in the matrix world, they have to be plugged in. But, once plugged in, then the action takes different shapes in the matrix world. The martial arts, "Kung Fu," battles with Smith in all his multiplications, with flying interchanges is the program world where the ordinary laws of physics are suspended. And it is the "world" about which we must most strenuously suspend our disbelief.

If, as I am contending, this tradition, Plato-Descartes/Locke-Matrix, some technology or technology complex provides the model for knowledge and human experience, then in its latest incarnation we can expect a differently shaped theatre, and I believe the late modern adaptation of computerization, situated in proximity with that most "Cartesian" of late modern sciences, neurology and some versions of cognitive science, give us the clue for the popularization of this epistemology engine in the *Matrix* series. Descartes, in his earlier version, was indeed puzzled how a mind – the homunculus in the camera – could be connected to a mechanical body, and he theorized several versions of a "connector." One was a distribution of mind throughout the body, but the other localized connector, he thought to be the *pituitary gland,* located in the middle of the brain. Today's version locates the connector in, or *as the brain*. But this mind is also different from Descartes. To-

## 2. Technofantasies and Embodiment 9

day's brain, the homunculus's new version, is now an autonomous "computer" or "hard-wired" brain which de-codes "information" which comes in via the various sense organs (eyes, ears, tongue, fingers, etc). And if this is so, there is only a very small step between this notion of brain and the possibility of the *Matrix*. Or, reversing the metaphor, the *Matrix* is a cinematographic version of the latest epistemology engine: inner brain processing interacting with external data-code input.

The*Matrix* trilogy is simply a contemporary variant upon a very ancient set of human imaginations, imaginations which combine embodiment fantasies with some form of materiality, frequently *technological materialities*. This is why I call them *technofantasies*. One such fantasy, ancient, multicultural, possibly universal, is the fantasy of *flight*. But the ways in which humans fantasize flight also differs: some versions of flight entail types of mysticism—one leaves one's actual body and is transported somewhere—heaven, hell, astral regions, wherever. But, in many cases flight is accomplished through some kind of agency. Animal agency is common: one flies upon the back of some large bird, or a flying dragon, or a chimera, flying horse, etc.etc. etc. And, in still other cases a 'technology': a flying carpet, a machine, or other technological device. For the Greeks there is the story of Daedalus: Daedalus, a mythical inventor, who makes wings from feathers and wax and although warned, his son, Icarus flies too close to the sun, melting the wax and he plunges to his death in the sea. Here is a technofantasy projection, with which both poor technologies and with no knowledge of aerodynamics, simply was not possible. The flying fantasy is clearly ancient and Icarus's technology employed a flight machine which could not work. Why such a technology? Although I shall not pursue this here, I suspect that assisted flight fantasies depend, in part, upon cultural lifeworld variants. Technology assisted flight—in contrast to psychic or animal assisted flight—may require greater familiarity with some greater machinic texture to a lifeworld. Lynn White, Jr., is famous for his histories of early technological developments in Europe from the very early through the high middle ages. The $13^{th}$ century was particularly prolific: mechanical clocks had spread to town halls, cathedrals and clock towers. Large windmills, water wheels and donkey driven cranes performed labour tasks which no longer depended upon human energy alone. Eyeglasses and other lens technologies were also widespread. Soon after cannons replaced catapults and castles began to fall. All this was part of a much more mechanically textured

## 2. Technofantasies and Embodiment

lifeworld already in place before the Renaissance. And, as this mechanical proliferation began to be part of common experience, it also entered human thinking as a kind of epistemology engine: as early as 1270, Roger Bacon, the first European to write an optics, also began to make fantasy projections of imaginable technologies. He wrote about flying machines, armoured and self-propelled military devices, underwater vessels and the like. Bacon's age had already begun to be filled with technologies which stimulated his imagination. White points out that by the time clocks and eyeglasses and large-geared technologies were part of daily experience, the mathematician-ecclesiastic, Nicholas Oresmes (d. 1382) took the metaphor of a clockwork and applied it to the universe itself. Leonardo da Vinci, by the mid 1400's was doing technical drawings, actually based on Bacon's verbal descriptions, of his odd screw-driven flying machine, many geared devices, a diving bell with a bubble-helmet (I have seen models of many of these in Vinci, Italy, his home town) and as all critical engineers know, virtually none would have worked! Just as wax and feathers, neither materials engineering, engine power, or aerodynamics was part of even this Renaissance scene. What I am hinting at, is that in an already technology-familiar culture, fantasies can easily take such technofantasy forms.

Later attempts at actual technology-assisted human flight remain familiar, for example in early movie documentaries with all the funny jump-and-crash scenes portrayed. These were clumsy attempts to fulfill the fantasy. but by mid-$20^{th}$ century new light weight materials and bicycle-like gearing showed greater promise. The '80's "Daedalus Project" using kevlar and mylar materials, an adjustable propeller bicycle gear driven and a trained athlete, both in 1987 and 1988, try to fly from Crete to Santornin, repeating Daedalus and Icarus. But minimal powered flight for a set of wings capable of supporting a human, calls for 10 horsepower and the athlete puts out 0.4 horsepower and becomes exhausted and crashes onto a beach. Later, 1996, competing for a $100,000 to cross the English Channel, in ideal conditions, another athlete succeeds in a similar light-weight, bicycle-like powered machine. No danger in any of these attempts of rising to the sun, and both barely make it, fully exhausted. But, it was a meeting of hi-tech materials, aerodynamic science, and ideal conditions for a "human powered" flight, fulfilled in some sense, but hardly exemplifying the ease imagined by technofantasies. Literally or minimally, the fantasy was fulfilled, yet it never reaches bird-like quality.

## 2. Technofantasies and Embodiment 11

Technofantasies include many sorts of desires, not only flying, but in general technologies which will give us *powers* usually beyond our bodily, sensory, sexual, intellectual, or for that matter any or all dimensions of human embodiment. But while we imagine technologies which could do this, we also want them to be transparent, without effort, enacted with ease, as if our enhancements were part of a well trained "sports body." And this is the juncture at which we can also return to the late modern technologies which seem to promise us the materialization of just such a fantasy.

I contend that the technologies which aid in *Matrix-like technofantasies*, are the modern and late modern *imaging technologies*. But they only produce an *image* and, going all the way back to Plato, I will argue they *do not embody*. The Matrix is a hi-tech cinematographic imaging process, bringing into play all sorts of new special effects, capturing our fantasy and attention. But the route to the *Matrix*, reflects this much longer history of imaging. Slightly modify the lens opening to the *obscura*, add a shutter, insert photographic, light sensitive film, miniaturize and you have a "Kodak." All this takes place from the mid-to-late $19^{th}$ century. This late modern imaging, however, "image" something very different from more ancient imaging and can be described as the technological manipulation of time-experience. Again a foreshortened history – Joseph Niepce (1826) managed to project images on light sensitive film, but with very long exposure limits thus also limiting the choice of objects to only stand-still objects, for example architecture; Louis Daguerre (1839) perfects and begins to speed up exposure time, with portraits – as long as you can stay very still – becoming popular. Eduard Muybridge (1878) made the leap to milliseconds with multiple cameras and did thousands of motion studies, galloping horses with all four feet off the ground, naked men and women in motion, now inventing time imaging manipulation which began to make what was never before *perceivable* visible in images. A French photographer did the same to show how a cat, dropped upside down, righted itself before landing on the ground. And, then, finally there is motion picture photography, the "movies," with the Lumière brothers, from 1895.

For anyone born after the 1920's, when sound is added to film – all this imaging, from stills to time lapse photography – allows those who *experience this style of imaging perceptually, audiovisually – for these, this imaging has become of the contemporary visual lifeworld*. But, it is a visualization with a difference; it is not

## 2. Technofantasies and Embodiment

fully embodied experience and it is what could be called a kind of *counterfactual visualization* – it is a display or "theatre." We do indeed see the isomorphic, photographic "realism" of Matthew Brady's Civil War photographs; we see earliest *cinéma vérité* shots of a train entering a station; we see the reversed motion shots of the cat jumping backward to its upside down position; and we see the time lapse photography of the sunflower following the sun through the day. Time stops, speeds up, reverses, slows down. Our whole bodies can see the visual spectacle, beyond ordinary vision, but our full embodiment does not participate. It remains constrained, despite desire, despite fantasy. No cat can reverse its jump; and were we to mimic the sunflower, standing in one place and keeping our face turned towards the sun motion, not only would this be difficult, but we would probably be so impatient as to give up the task. Or what about the astronomer who today can actually produce emission images of 13 billion old galaxies? Even in spite of the style of mathematical physics which sees no reason for non-reversible time, no one has ever imaged a 13 billion old *future* galaxy. Galaxy realism is always limited to its past emissions.

In technofantasies, constraints tend to disappear. Flying, in the *Matrix,* is not technology assisted. Neo and Agent Smith, using the visual quotations from earlier Kung Fu movies, fly at each other. This tradition of non-assisted human flight may derive from the earlier fantasies associated with "Superman." Neo learns to speed up and slow down his motions such that he stops bullets which are imaged like the fast arrows of "Men in Tights," and there is even time reversal with saving Trinity. From our ordinary, now imaged world, this is visualized. But where are *we*? Like Plato's prisoners, or like Descartes' homunculus, we remain in our forward-oriented, relaxed bodily position, *inside the theatre.* This is not like a rock concert where we are getting into the music we embody; not reversible, but hypnotic whole-body motion, a bodily motion which moves with the music, by all actors in the same context. The rock star can be thrown out to land on the upturned hands of the audience; but in cinema, we do not jump up on stage to punch out Agent Smith. I am suggesting that there is a unique kind of embodiment which allows the visual fantasy enhancement of something like a *Matrix* technofantasy to be what it is.

Early imagers recognized this problem and tried to add more complete sensory experience. I have referred to IMAX with its 70mm projection and surround sound; other versions have added

## 2. Technofantasies and Embodiment 13

shaking seats and floors, or, as in airplane simulators, motions added to the cave-analogues. There was even "smellovision." These are the technological trajectories aimed toward today's virtual and augmented realities. But are these in fact "better"? In early 2004, I played inside two "virtual reality caves" in Umea, Sweden. Goggles, 3-d, and a hand-held pointer, my motions were monitored by ceiling sensors; the 3-d projections were co-ordinated with my bodily motion and I could choose where I would go, following a sort of video-game like context. Previous unreal or odd senses of bodily motion occurred again, but this time I decided to seek the program's constraints. I tried to penetrate the wall of the castle, but I could not; I tried to jump off the cliff and fly over the sea, but I could not; what I could do; I was limited to following the game plot. This experience was one in which my sense of embodiment produced a stronger contrast with the fantasy of the game than any cinema experience. To me, this was even more obvious than the chained restraint system of Plato's cave. The *Matrix*, of course, does better, probably in part because it is visualized rather than embodied. Neo learns to stretch the program with his time stops, reversals and speed ups, in addition to his Kung Fu movements; Agent Smith learns, like a virus or worm, to replicate himself and develop almost a personality of his own inside the program. Embodiment, I hold, contains the clues to recognizing constraints and thus the plays upon appearance/reality, whether in the cave, the *camera*, or the *Matrix*. In short, I remain a phenomenological skeptic with regard to this entire tradition of technofantasy and embodiment. I cannot imagine any of the variants of entering a theatre and yet not knowing one is entering a theatre with its demand of a suspension of belief; only the oaf rushes onto the stage and beats up the villain who demands the rent from the hapless victim of a melodrama.

Is Moravec an oaf, then? And will he offer his mind for downloading into a computer program? It is at this point that I want to suggest, but only begin to suggest, a radically different interpretation of technologies as they relate to our lifeworld. Computer-enhanced, computer-processed, computer-tomographic processes are the latest versions of imaging technologies. They become more and more sophisticated, but remain short of experienced embodiment. From the rather limited "Agent Smiths" in the first *Matrix* film, with actors all made up to look like Smith, we move, by the third film, to the digitally imaged, "thousand" identical Smith replicants. Computer imaging, modeling, simulations, al-

## 2. Technofantasies and Embodiment

ready began to take shape in the mid-$20^{th}$ century. Today, any version of whole earth or environmental modeling, CT, computer-tomographic processes, *constructs images.* There are several important techniques involved, but one is especially relevant to Moravec's technofantasy: computers can convert digital codes into images and reverse the project. Not too many years ago, on my home computer I received an email attachment from my oldest son – I pulled it up and pressed "print." Many pages later, my printer produced some 24 pages of code; no one could tell what it was supposed to "mean." So, I asked my wife, Linda, to look at it at school on her optics and, in one page, the result was a digital photo of my son, his wife, and our new grandson. This was code-to-image and back again. Space probes, modeling, medical imaging, all include this technology. It is analogous to early photography, then movies, stop-time, speed up time, reverse time – *in imaging.* The *Matrix* actually forefronts this reversibility—the cascades of data displayed are the flip side of the Kung Fu flights. In short, once again, this imaging pattern has become part of our visual experience, within our lifeworld. It *seems* only a short jump to falling into our movies, our televisions, our imaging devices and thus supporting the Moravec fantasy. But is this really the case?

What if we give up our slippery slope, sliding into and out of technologies themselves and instead *embody ourselves through* our technologies? In his *Phenomenology of Perception,* Maurice Merleau-Ponty describes a blind man using a cane. He *experiences touch, feel, extension at the end of the cane.* His experience is directed through the cane and touches the sidewalk. This touch-at-a-distance has become well known to cognitive scientists and neurologists. The fact, for example, that through a prosthetic third arm Stelarc, the artist, is able to write, is noted by Andy Clark (2003). Clark notes that our "technologies become transparent" (I would prefer quasi-transparent, since there remains an "echo" awareness of holding the cane in the background). But strictly speaking this has always been the case. It is just that our fantasies did not turn out at all the way we expected them to be. While no human-invented airplane ever came near the performance of a humming bird or even a beetle, much less a red-tailed hawk, the stunt bi-plane I once flew in as a passenger, could fly straight up, flip over and dive and at speeds faster than any harrier. And, were I to be able to afford $30,000, I could myself fly an MIG-19 at supersonic speed, embodying the craft quasi-transparently as I make a banked turn – more complex than a cane, but embodied nevertheless. Yet, when I leave the plane on the runway, it remains, as it were, my past exo-skeleton in its parking spot. It even looks like, by placing an electrode saturated cap on my head, that I may be able to move a cursor at a distance by directing my own bodily electricity and experience a new style of embodiment through a technology.

So, let me conclude with something of a surprise: Moravec is looking in the wrong direction with his hoped for download of his mind into a computer. I, at least, have been going through this process for decades. Here I am, "in" my words, right "before" you; you can pick me up and read me anytime. The catch, of course, is not unlike the fulfillment of human flight, it will be disappointing because no matter how many times you come back to this essay, it repeats itself. It will not pop up as a new and different article before you each time. This essay may change in meaning in very different ways but hardly so "in my daily embodiment where I am my body," and not in "real time" – where I escape the download, always exceeding it, getting out of the cave, turning around inside the camera, or walking outside the theatre. As Merleau-Ponty said, "there is no inner man, man is in the world, and only in the world does he know himself" (1962: xi). In short, we do not need technofantasy to be technologically embodied; we need, instead, to develop the skills and imaginations to be creative through our technologies. Neo needs to 'unplug,' not to rid himself of technologies, but to remove the illusion that he cannot tell when he is entering or not entering a *theatre*.

**Works Cited:**

Clark, Andy. 2003. *Natural Born Cyborgs*. Oxford: Oxford University Press.

Ihde, Don. 1973. *Sense and Significance*. Pittsburgh: Duquesne University Press.

Merleau-Ponty, Maurice. 1962. *The Phenomenology of Perception*. Trans. C. Smith. London: Routledge and Kegan Paul.

# 3

# Technologies—Musics—Embodiments

*Today recorded music probably accounts for the single largest category of music listening. This essay seeks to re-frame the usual understanding of the role of that type of music. Here the history and phenomenology of instrumentally mediated musics examines pre-historic instruments and their relationship to skilled, embodied performance, to innovations in technologies which produce multistable trajectories which result in different musics. The ancient relationship between the technologies of archery and that of stringed instruments is both historically and phenomenologically examined. This narrative is then paralleled by a similar examination of the history and variations upon recorded and then electronically produced music. The interrelation of music-technologies-and embodiment underlies this interpretation of musical production.*

Michel Foucault, in trying to convince us that pre-modernity had a form of knowledge, an episteme, which is now past and no longer makes sense, claims that the symmetry of resemblances which ruled the sixteenth century led to a conclusion that "there are the same number of fishes in the water as there are animals . . . .the same number of beings in the water and the surface of the earth as there are in the sky, the inhabitants of the former corresponding to the latter..."[1]

We laugh or are amused—what sort of claim is this? Is it empirical? But if so, then, theoretically, we could go about confirming or disconfirming it by a count. Yet, what agency—the National Science Foundation, NASA—-would support a grant to do such a count? We know from the outset that such agencies would not, although in today's episteme, one might be able to fund a census which would take a count of polar bears over several years to determine if they are entering endangered species levels, or do a census of whales to see if they have recovered their populations enough to be hunted. Foucault's point is that the thinking which relies upon such presumed symmetries simply no longer has any bite; its episteme is dead.

The purpose of this essay is to undertake philosophical reflections upon recorded music, or as I prefer in a parallel to Foucault's epistemes, musics in the plural. I will begin my reflection with an attempt to locate us with respect to the many musics we may experience contemporarily and hint at something like the suggested census noted above. I shall first fore-front the listener, the human who hears or listens to musics. But in parallel fashion, I shall also place the listener into a context where the technologies which mediate the musics are also brought under scrutiny. Clearly, there are indefinitely large numbers of such musics which could be listened to: performed musics—chamber, classical, rock, ancient music consorts, street performers and the list expands and expands. Similarly, whatever types of performances might be listened to, such musics could also be recorded and thus listened to in the form of recorded music. Recorded musics are—we usually think—replicated musics. But here one must then also account for the plurality of recording technologies: iPods, Walkman, vinyl, CD, digital tape, Musak, radio, and again an indefinite list of technologies which mediate and present the recorded musics grows.

Yet, with an echo of the difficulty of determining if the number of birds equals the number of fishes, there is difficulty in determining how many listeners hear how many songs in how many ways. Yet there may be clues: Gold records are those which sell 500,000 per run; Platinum records are those which sell 1,000,000 per run. In 2006 in the US, there were 30 gold and 16 platinum runs, thus equaling 31 million records.[2] Now, how many people listened to those 31 million records and how many times was each song heard? And don't forget to count all the downloads which also occurred in the same year from the multiple sources. While we do not know an actual number, we can easily surmise that the number must be very, very large indeed! But, to make it simpler, while yet remaining intuitive, let us imagine only listeners to recorded songs in greater New York on a given day, and then imagine the listeners to all the live music performances of whatever kind on that same day—from the philharmonic in Lincoln Center to the Peruvian flutists in Greenwich Village. I am willing to wager that the number of recorded presentations is simply much larger than the number heard in live performances by a very large magnitude. Here my point is that we philosophers, musicologists, performers and other theorists may be coming to reflect upon recorded music very late. For if it is the case, as I suspect, that on a world-wide basis more listeners listen to recorded music than any other kind of presented music, it is sort of like a very large canary already

## 3. Technologies—Musics—Embodiments

escaped from its cage and now has grown too big ever to re-enter. Recorded musics are, of course, technologically mediated musics. And, historically, recorded musics are quite recent arrivals upon the very ancient histories and even pre-histories of other musical presentations. For the moment I shall not discuss scoring, which could be considered a sort of pre-recording technique for preserving some kind of identity between the same piece played in different performances, and which also employs a type of 'material technology' analogous to print, but as notations on a page. Nor shall I do more than sketch the very rapid history of recorded musics which go back only 130 years, but I do want to point up in bulleted blinks, how fast and how diverse this technological trajectory has been:

- 1877 Thomas Edison produces the first useable cylinder recording, first with tinfoil, later wax cylinders. These were mechanical devices recording sound waves physically, mechanically. Only a few plays are possible before the record deteriorates.

- By 1899 coin-in-box predecessors to juke boxes were already popular.

- By 1902 Caruso began to record, first with cylinders, later with discs, which began to appear in 1903.

- 1904 saw the invention of the diode which made electrical rather than mechanical recording possible, but which did not become practical until 1919.

- By 1923 radio threatened to depress the reproduction industries, first with live, later with recorded presentations.

- The first stereo developments began in 1931 and magnetic tape followed in 1934.

- Vinyl, which reduced surface noise compared to the older discs, began in 1948 and the older '78's began to be replaced by '45's and 33's. Full stereo was available by 1956.

- Cassettes, 1963, digital CDs, 1978, and DAT or digital tapes came on in 1989.

- Then, with the 1990's came the proliferation of online and download-ing copies in all the varieties now popular.

20        3. Technologies—Musics—Embodiments

- I want here simply to make two points: first, this 130 year proliferation must appear as a very rapid proliferation which also covers a wide variety of different technologies to record and reproduce musics. Second, it is a history, which while having ups and downs, clearly is one which now pervades an entire global economy, again evidence of our very large escaped canary.

I have now begun with technologies, which in the case of recording technologies mediate musics which in this first pass are heard by listeners. In short, I am relating here a material means of producing musics to experiencing humans who listen to these musical phenomena. Now, however, I want to shift to pre-history and begin now a long-range location for musics. How long ago humans began to make music remains unknown—but whenever music began, from the earliest human beginnings there were always already present human uses of technologies! This may seem like a strange claim, but only a few years ago there appeared in *Science* an article which refers to the chipped stone tools used by chimpanzees to crack nuts in a hammer/anvil fashion which go back at least 4300 years BP [before present][3] The sample found remains quite identical with contemporary chimp tool formation and use. The surmise of the article is that such simple tool use probably goes back at least to the common ancestor from which chimpanzees and humans split off! By current dating that is over six million years ago. And, for the physical anthropology literate, we are all familiar with tool uses by homo erectus, Neanderthals, and homo sapiens sapiens. Stone Age tools go back at least two plus million years. I would like to suggest that our common image here is one of a limited appreciation of the diversity of technologies by our ancient ancestors—we may think of Acheulean hand axes, or chipping tools, or maybe if we realize that 'soft' technologies such as nets and baskets probably were also used, but all of these, we usually surmise, are for subsistence needs. That is, we tend to evoke a simple and basic existence for our predecessors. This may underestimate our ancestors.

What, then, should we make of a 45,000 BP "bear bone flute" found in a site associated with Neanderthal humans in a cave in Slovenia? *Science* reported this find and the probable conclusion that this artifact was likely a flute as evidenced by the regular symmetrical shaping of the four holes which, under analysis, yield a tuning system for a diatonic scale.[4] And while such a musical instrument is not millions of years old, so as to compete with Acheulean hand axes, it probably does suggest that performed instrumental music occurred long ago in pre-history. In passing, note that we are now shifting the music scene from listeners—although they, too, were likely present—to also include performers. With performers, human embodiment actions come into play. The flute player must learn an embodiment skill which engages, in this case, the disciplined hand and breath motions which are mediated through the flute to produce music. We are now able to recognize a very basic relational ontology of instrument use. The human practicioner plays the flute to produce musical sound—I diagram this as:

Human → flute → music

This relational phenomenon is what phenomenologists call 'intentionality.' But in this case, it is an actional intentionality which is directed, mediated through a material instrument—a technology. A deeper analysis would go on to show that in the learning process, the shapes of experience change: first, struggles with playing the flute yield sounds, but they are not refined, gracile, 'musical.' But as skill is acquired, the flute is 'mastered' in that it withdraws or becomes more and more transparent and the player is able to produce the sounds we hear as flute-music. This same process, which we can describe for the movement from novice to virtuouso performance, no doubt also characterized the experience and attainment of the Neanderthal flutist. The 'woodwind,' here 'bonewind,' instrument permits the mediation of the human hand and breathing action into flute music, hearable both by the player and any audience present.

We have now taken a simple look at recent recording technologies and then at a very ancient instrumental technology with musics mediated by different technologies over a very vast historical span. Now I want to risk reader dizziness by reciting what for me was a very formative occasion, a conference on musical improvisation at the University of California, San Diego, in 1981: I had been invited to be a keynote speaker at this conference which appeared to me to be of great interest—an interdisciplinary gathering of musicians, composers, humanities academics and even one other philosopher, Daniel Charles from the university of Paris. I arrived, realizing that I knew not one person from previous experience, although at social events I met folk who had read my *Listening and Voice: A Phenomenology of Sound* which was, of course, the connection to this event and the source which motivated the invitation. But it was the improvisation workshops which turned out to produce the most interesting provocations.

I arrived at a workshop, a studio in which there was a grand piano, various traditional instruments, and a collection of distinctly non-traditional instruments. I tried to be unobtrusive and found a seat in a corner, self-consciously aware that my only performance abilities were abandoned long ago from high school to early undergraduate trombone playing days. But observer status was not an option—I was handed a "water horn" and commanded to participate. The water horn was a stainless steel container, partially filled with water, to which had been brazed a series of brass rods of different lengths around the perimeter of the vessel. I was handed a violin bow, and by now cacophony had already begun. Some players were hitting the open piano wires with hammers; others were turning anything in sight into a percussion instrument; virtually nothing was being played in a standard or traditional way and so, in the spirit of the event—which was simultaneously being recorded for posterity—I began to bow the rods on my water horn, and later to shake the instrument to get a gurgling sound. Extreme improvisation indeed, but this event took in my experience and memory as I realized that any instrument whatsoever had multistable possibilities just as I had earlier noted belonged to various other perceptual phenomena. This event, 1981, occurred not long after my first book on the philosophy of technology (*Technics and Praxis*, 1979) in which I had initiated a long term interest in the role of instruments as the material means for producing scientific knowledge. And, I claim, there is a parallel between scientific and musical instruments with respect to the already noted role of intentionality. The human action undertaken is mediated through the instrument to produce its result. If, in one case, it is the greater knowledge Galileo obtained through his telescope—of the craters of the Moon, the satellites of Jupiter, the phases of Venus, in the other case it is the transformed sounds produced through the playing of the instrument, sounds which are as different from ordinary human vocal sounds as the sights through the telescope differ from those of naked eye observation. And to attain each production, in both cases, there is an acquisition of a skill which must be acquired to have a refined and high quality result. I shall return to this parallelism later for added analysis, but for now I want to return once again to an ancient pre-historical ex-

3. Technologies—Musics—Embodiments    23

ample of technologies which turn out to display a remarkable set of multistabilities which I find surprising. My example is that of ancient archery: with few exceptions [Australian Aboriginals who developed boomerang technologies and some southern hemisphere groups who developed blow gun technologies] virtually every ancient culture developed archery, variations upon bows and arrows. But the style of use, the technical composition of the material artifacts, and the cultural contexts into which these technologies fit, display an amazing set of multistable patterns:

- The English longbow, constructed of yew or ash, long (2 M.+) with long arrows, could be taught simply for use for the yeoman and fired with rapidity from a standing position. The firing technique called for the bow to be held at arm's length before the bowman, with the bowstring then pulled back and the arrow placed between the first and second fingers for release. Its effectiveness was demonstrated in the historical battle of Agincourt, where this type of archery overcame the crossbow archery of the enemy.

- From the East came the radically re-curved, short (1.2-3 M) and composite bow used by Mongol horsemen who repeatedly invaded eastern Europe. Clearly a bow of longbow length could not work on a galloping horse, and the composite of bone, wood, skin and glue, radically recurved, allowed a smaller weapon similar power. But the firing technique was also different. Here the bowstring is held close to the face and bow pushed rapidly away with another form of quick fire, timed exactly to the gallop of the horse.

- A third 'artillary' style of archery arose in China, in this case a long (2M+), recurved bow called for the highest pull power in antiquity (matched only today with the compound pulley powered bows now popular). Here the firing technique included simultaneous push and pull, plus the use of a thumb ring to prevent injury to the thumb by the string. I was delighted to discover some of the terra cotta warriors in the XiAn complex cast in exactly this posture on my first visit to China in 2004!

The examples above illustrate what I call multistability in the sense that the 'same' technology takes quite different shapes in different contexts. In each case, the tensioned, strung bow can propel the arrow over distance with striking power, but the human skills take different shapes and patterns in using the also technically different artifacts. Does this seeming excursus have anything to do with music? My answer is 'yes' and I shall try to open this line of inquiry by returning to the improvisation event with its playful exploration of performance variations upon traditional and nontraditional instruments.

Every archer could hear the bow string 'twang' when fired. Could it then be 'played?' We have already noted at least three styles of firing an arrow: bow extended and held still; string held still and bow pushed out; and double push and pull. Each of these variations, however, serve the same purpose, to fire an arrow. But in a new context if one holds the bow in a horizontal position instead, and 'plucks' the bowstring—we are transforming the bow from its usual use, into a new use, as a sort of stringed instrument!

Anyone familiar with a history of instruments, or ethnomusicologists would know that there have been a wide variety of single-string instruments in many cultures. But I suspect not many know that there is also a tradition of actually using an archery bow as a stringed instrument. Apparently such uses are common in Africa. According to even such a common source as the Encyclopedia Britannica, "...The San of the Kalahari often convert their hunting bows to musical use."[5] So, we are now off on a new technological trajectory or line of development. To produce a more interesting music, why not a moving fret? And, we notice, that the simple 'twang' is not very loud or powerful, so why not add a resonator?

Again, just as with the hunting bow, there are a variety of types: "There are three types: bows with a separate resonator; bows with attached resonators; and mouth bows [all] evolved from the hunting bow."[6] In short, what I am doing with this set of cultural variations, is to show how an improviser creates a new stability or performance practice, using either a regular archery bow, or by adding very minor features and thus open-ing the way to developing stringed instruments. In my actual, original speculations—phenomenological fantasy variations—I did not yet know about what I cited above, but followed what I knew to be an exploration of different ways in which the human-technology actions could be possible as in the improvisation event. But as it turns out, this trajectory had already been followed. The above illustrations are of sub-Saharan African practices. But such practices apparently are not only contemporary practices. Another source, archeologist, J.D. Lewis-Williams points out in the *South African Archeological*

## 3. Technologies—Musics—Embodiments 25

*Society Newsletter*:
...Bushmen recognize two kinds of music, vocal music and instrumental music. ....the more personal instrumental music ...is played ...sometimes while walking in the veld, sometimes while relaxing in the camp. One of the most characteristic Bushman instruments is the musical bow, which is, in fact, an ordinary hunting bow. It can be played with the performer using his mouth as a resonator. When so used, the bow is more or less horizontal but can also be played with the stave vertical or semi-vertical. Then some performers like to use a calabash or other object as a resonator to produce more varied sounds as they tap the string with a stick.[7]

Then, to clinch the case, these modes of playing the bow have recently also been found to be depicted in the traditional Bushman rock art found in the Natal Drakensberg area by Paul den Hoed and Justin Clarke:

Bushman rock art found in the Natal Drakensberg Reproduction by Don Ihde

This rock art dates back to 2500 BP. (I visited this area and saw some of this rock art in 1982 but did not see this particular depiction.). And, if this is not enough, note that a shaman depicted in the Trois Freres cave in Southern France with a dating of 15,000 BP is shown, as now commonly interpreted as playing a bow in this same position (see next page).

26   3. Technologies—Musics—Embodiments

With bows played in this way, both multistable variations appear, and in each different types of musics are produced. "Apart from adapted shoot-ing bows, more specialized types of musical bows are widespread. Most are sounded by plucking or striking the string, but the Xhosa uubhu is bowed with a friction stick..."[8] Thus the 'same technology'—a bow—apparently fits two radically different trajectories, one of them musical. And this set of different trajectories is apparently also very ancient.

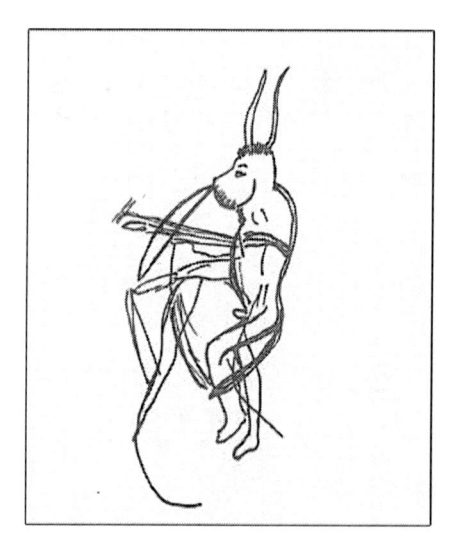

Bow playing Shaman, Trois Freres Cave, France Reproduction by Don Ihde

On the surface, it may seem that this detour into pre-history with glimpses of ancient 'bearwind'/woodwind instruments, or of stringed weapons/stringed instruments may seem very far from the focus upon recorded music. But my point here is that, in human-techology interrelations, there may be found a set of multistable variations, and, from these, suggestions for trajectories or further and different developments. To fortify both points permit one more set of variations, this time those which may be found in a similar comparison between scientific and musical instruments.

It is probably not coincidental that the European Renaissance and Early Modern Science both marked a period in which instrumentation began to proliferate in both art and science practice.[9]

### 3. Technologies—Musics—Embodiments 27

In music, this is a period which instruments are more and more used, compared to older a cappella and plainsong sacred music, to the increased use and experimentation with a variety of stringed, brass, woodwind and percussion instruments. Indeed, our current orchestral instruments such as violins, violas, horns, timpani and the like, can all trace their development to this period. Musical instruments, however, produce sounds but many other Renaissance and early modern instruments were more related to optics and visualizations. Galileo, often taken as the paradigmatic figure for early modern science, developed both telescopes and microscopes utilizing compound lenses to magnify both the macroscopic and the microscopic phenomena of interest. I have referred above to several of his observations of previously unseen celestial phenomena—and these could only become visible by first, recognizing the optical possibility of magnification, and then gradually developing the optical technology which both allowed greater resolution and greater magnification. Indeed, Galileo, hearing of Lippershey's 3X telescope, went on to produce nearly a hundred of his own telescopes, up to approximately 30X, which turns out to be the limit for lenses without encountering chromatic distortion.[10] Thus, the limit Galileo reached was one which allowed him to recognize the "protuberances" of Saturn, but not resolve them into the rings with which we are now familiar. The trajectory, of course, is the line of development which recognizes in the material possibilities of optics, the possibility of greater and greater magnification and resolution, suggested in the very use of the instrument. The same following of a trajectory is, of course, also possible with musical instruments. Changes of material for stringed instruments, for example from gut to hair to wire or polymer strings, all allow different tonalities for the produced sounds. Nor should we forget that the human actions in 'playing' or tuning both scientific and musical instruments also plays a crucial role in the output. Galileo insisted that in order to see what he saw, he needed to train the novice telescope user, not unlike the training which must go into producing a good tone and sound from any musical instrument.

There is, however, a crucial difference which may illustrate a quite different sub-cultural contrast between the history of scientific, compared to musical instrumentation. Once again, I revert to a somewhat imaginative variation for illustration: I ask—could you imagine a serious $21^{st}$ century astronomer directing the graduate students in astronomy to try to make a new discovery of some astronomical phenomenon by picking up and re-using one of

Galileo's telescopes? Yet, especially when I visit Europe, I have often been delighted to attend concerts given by groups who pick up and use ancient or early instruments to perform a chamber piece! And, even today, what violinist would turn down the opportunity to play a concerto using a Stradivarius or Guaneri violin? I am suggesting that there is lurking here a strong contrast between the instrumental traditions of much science and of much music culture. In general, I am claiming that there is an inbuilt progressivism in the adaptation to new instrumentation associated with science practice, but there can be an equally inbuilt romanticism which sometimes results in a preferred traditionalism associated with some music culture. This may also be evidenced by historical examples.

For example, Trevor Pinch and Karin Bijsterveld have long been interested in the way in which new technologies have been received within historical musical culture, and they have discovered what I shall call "Heideggerian moments." They note in "Breaches and Boundaries in the Reception of New Technology in Music," that technological innovations are frequently first decried, "For instance, the introduction of the piano forte was seen by some as an unwarranted intrusion of a mechanical device into a musical culture which revered the harpsichord."[11] And, again, when in the $19^{th}$ century, "...key mechanisms and valves (such as found on today's woodwind instruments) were introduced to replace the traditional means of controlling pitch by the use of fingers over the individual holes. The new valves and keys were found to be easy to operate and facilitated the produc-tion of much more uniform and cleaner tones for individual notes [such a change met opposition from one, Heinrich Grenser, who complained that improving tone... by the use of keys was ] ... Neither complex nor art... the real art of flute construction was to build flutes which would enable flutists to play whatever they wanted without the use of keys."[12] I call this a "Heideggerian moment" because the objection to mechanical valves and keys parallels Heidegger's famous rejection of typewriters in favor of the pen:

Human beings "act" through the hand; for the hand is, like the word, a distinguishing characteristic of humans. Only a being, such as the human, that "has" the word can and must "have hands." . . . the hand contains the essence of the human being because the word, as the essential region of the hand, is the essential ground of being human.

[After which Heidegger goes on to discredit the typewriter]... It is not by chance that modern man writes "with" the typewriter and "dictates...into" the machine. This "history of the kinds of writing is at the same time one of the major reaons for the increasing destruction of the word. The word no longer passes through the hand as it writes and acts authentically but through the mechanized pressure of the hand. The typewriter snatches script from the essential realm of the hand—and this means the hand is removed from the essential realm of the word.[13]

One can easily substitute hand harp playing for keyboard piano playing and one can see the equivalence to the "Heidegger moment" in the resistance to changes in music technologies.

This resistance is perhaps understandable in the following sense. I have noted above that with any musical instrument which entails human bodily action and the acquistion of skills, presupposes long practice and develop-ment which the introduction of a new technology may disrupt. Moreover, any new technological development also enhances and simultaneously reduces some quality to the produced sounds which also thus changes the music. Yet, at the same time, new skills can be acquired and a new virtuosity may be attained. To those who complained about the 'mechanical' output of the piano forte, who today would think of Ashkenaszy or Ash as producing 'mechanical' sounds from the piano? It would be very hard for one who had spent years of hours of skill acquisition to simply abandon such skills for every new modification in instrumentation.

Only now am I ready to return to the focal topic of recorded music. Take note of the implicit trajectory in the short history of recorded music: the earliest cylinder records had very poor fidelity, could play only very short pieces, and could replay them only a few times. In spite of this early listeners often gasped at how "realistic" the voices sounded. Yet, in the interplay of designers and listeners to this recorded music, it could be immediately clear that, if it was possible to improve fidelity, lengthen playing time, and increase repeatability, one should do so. and the re-examination of the noted history shows that this aim was followed:

- Most of the early cylinders allowed only 2.5 to 3 minutes of playing time; this led to matching the musical piece to the recording time capacity and 'popular' tunes of segments of arias prevailed.

- By 1903 discs had begun to prevail, but still playing time was short thus HMV Italiana's release of Verdi's "Emani"

30    3. Technologies—Musics—Embodiments

took 40 discs.

- The leap from mechanical to electrical recording, 1877 to 1919, was irreversible, but background noise still posed a fidelity problem.

- Stereo, initiated in 1931, added depth to recordings and vinyl, 1948, combined to produce much higher quality recordings.

- The switch from analog to digital and the invention of the CD, paral-leled by the amplifier tube to transister technologies produced changes in sound qualities which found loyalists who divided between listeners who preferred the richer tonalities of vinyl and tape recordings over the background noiseless, but crisper digital sounds.

This history, shorter than that of early modern optics, displays the same trajectory desires, here for greater sound fidelity parallel to the visual desire for greater magnification and resolution. In the process, however, something else also happens: the technologies for producing these results become much more complex and compounded and with this evolution there arises the possibility of greater manipulability. In order to return to recorded music technologies with a new perspective, I will here return to the music parallel and my final science instrument examples.

Early modern astronomy experimented with compound lens telescopes, all of the refracting sort (tubes with lenses and a focusing device), but, as mentioned, by the time one reaches 30X the fact that white light is made up of a spectrum of colors which refract at slightly different wave frequencies, meant that resolution would be poor and a chromatic distortion sets in. Newton, a century after Galileo, discovered this and reasoned that if one could use a parabolic mirror to re-focus the different wave lengths, one could overcome the chromatic distortion—thus the reflecting telescope which combines lenses and mirrors. Then also, as multistable variants were tried with optics—prisms instead of lenses producing spectra, twin slits producing wave imaging—more and more compound devices produced more and more previously unknown phenomena. A truly revolutionary step was taken, however, only with the discovery of digital processing utilizing computers in the $20^{th}$ century, which made possible many of the images now familiar within astronomy. Computer tomography makes possible the vari-ous manipulations which today show us everything from

3. Technologies—Musics—Embodiments    31

extra-solar planets to spinning pulsars.[14] Indeed, I have a close astronomer friend who claims that a telescope is no longer even considered an instrument; it is merely a light gathering device to which are attached the variety of 'instruments' such as spectrometers, interferometers, etc.

I am now finally ready to return to recorded music and its technolo-gies, but with a new framework for understanding, a framework which is no longer bound to taking recorded music as simply copies of, or representations of previous or simply performed music! Rather—and this is the crucial shift—the new combinations of technologies in a complex gestalt could themselves be considered to be a different instrument or instrument set. In my simplest examples above, such as adding a gourd resonator to a hunting bow, one changes and makes the instrument both more complex and yet more 'resonant' for the musical sound produced. Admittedly, the much later addition of electronic amplification is a more dramatic change but one would not deny that the electric guitars played by rock musicians are simply different instruments—indeed one can hardly imagine a rock concert without extensive systems of amplification. And more subtle forms of amplification, often subtly 'hidden' now, are part of the musics of opera, Broadway shows, and other performances.

In this re-framing of the understanding of how musics are technologically mediated, add once again a possible improvisation trajectory to what seems to be recorded music. Here I now take note of transforming what is initially recorded music in a 'performed' or constructed music direction:

- Perhaps the simplest and most direct transformation of recording technologies into instrumental performance ones is the "D.J." use of records being played, which are then, hands-on, manipulated by the DJ thus changing the sounds. In this case the 'same' technology which produced a music for 'passive' listening, is changed into a transformed music.

- Second, and drawing from the above history of recording technologies, a more deliberate and creative transformation comes from a $20^{th}$ century example—the music of Gyorgi Legeti. Here, he utilized a cut and paste, or bricollage construction of a composition is made from previously recorded sound bits to produce entirely new and different music. The end result is nothing like the previous musics which were recorded, but is a new music with its own gestalt sonic char-

32   3. Technologies—Musics—Embodiments

acter.

- In both the previous examples, the musics mediated by recording technologies are produced by rearrangements and reconstructions.

A third transformation with deeper implications relates to the initial production of recorded musics. The emergence of the sound studio which bears a direct parallelism with a science laboratory, provides the possibility for further manipulation and transformation of sounds. The sound studio technician—like the lab technician—tunes and tweaks the sounds on the way to being recorded. My point here is that one needs to envision here the much more 'corporate' or team involved in music production. It is not simply the individual or group playing which makes the music; it is rather the whole complex of processes, persons and technologies which produce the music.

- Then, as previously noted from our recording technology history, this growing aggregation of parts from which music is produced, is largely a $20^{th}$ century phenomenon with all of the above getting as far as a movement from mechanical to electronic mediating technologies. With the late $20^{th}$-into-$21^{st}$ century development we reach the level of digital and computer assisted sound producing technologies. This development signals a final break from the implicit 'copy' or 're-produce' model of sound production and shifts to synthesized, generated sound which is no longer necessarily based upon copied or recorded sounds—digital-computerized music does not need an 'original,' but is itself an 'original.' Today, of course, there are many examples of such musics, ranging from techno, to electronic synthesized, to totally transformed musics mediated by computerized-digital technologies. One innovative example comes from the work of Felix Hess, a physicist-turned-artist. His book, *Light as Air*, containing of course a CD of his produced musics, takes recording into a different style of construction. Admittedly, he often draws from 'natural' sounds in the sense that one piece uses sensors which use the window panes of apartment complexes as speaker diaphragms. The panes vibrate when sound is produced within the room and the sensors pick up this sound and 'record' it. But then the sounds thus collected over possibly days, is computer tomographically time

### 3. Technologies—Musics—Embodiments 33

compressed so that a 24 hour period is reduced to 8 minutes of played sound, a new music.[15]

What I am suggesting is that the boundaries between recorded music and a new, complex and technologically developing music production are blurred. It is here that I can return to my opening with a Foucault-like episteme. What I discern in the histories I have traced, is that musics which include instrumental technologies, may now be re-conceived and understood in many ways as parallel to what has happened in science instrumentation. The old image or episteme of early modern science was one which romanticized the genius individual—Galileo, Leewenhoek—each producing his own instruments, telescopes and microscopes, and bravely discovering through newly mediated observation, the new phenomena which made early modern science what it was. This is surely not the episteme of late modern, or possibly postmodern science. Rather, the new instruments are the large, complex colliders and particle collectors, which are 'played' and tuned by many, many scientists and technicians in the Big Science of today.

Music, in a different but often parallel fashion, and perhaps most clearly seen under my image of the large escaped canary of recording technologies, shows precisely the same kind of shift. But we have rarely recognized that instruments can be not only simple, relatively small and perhaps individually played, as with my gourd bow example, but they can also be large, complex, high tech and communal in the production of new musics.

This, in turn, casts a different perspective upon recorded music. Much recorded music today is a sort of doubled reproduction. Recorded results of studio produced and manipulated musics, are pretty much what books are to textual productions. They are the 'calcified, materialized' artifacts which we can pick up and read or re-read at will and leisure. I could have begun with the book metaphor, of course, but ending with it perhaps helps with understanding some of the ambivalence so many music theorists feel about recorded musics. When reframed by a book metaphor, while there remains vestigially something of 'copy' notions of an original, such 'copy' and 'reproduction' notions become weaker. No one expects readers of books to regret not having an original hand written manuscript instead of a nicely printed book. And such a reframing also should weaken any elitism which often gets expressed in disdain for recorded music from music theorists. In parallel, from a humanities perspective, I doubt most humanists would ever think of decrying the world of books as texts in the way in which some music theorists decry their auditory counterparts, recorded music. Neither the book, nor the record is dead. And both are only variants in the wide, wide world of language and musics.

## Endnotes:

1. Michel Foucault, *The Order of Things: An Archeology of the Human Sciences* (New York: Vintage books, 1973). Foucault is citing T. Camenella *Realis Philosophia* 1623, p. 98.

2. I am here following the numbers of the US standard; those of the UK area on a smaller scale.

3. Mercater, et.al., Science, Vol. 216, No. 5572, 24 May 2002, p. 1380.

4. Science, Vol. 291, No. 5501, 5 January 2001. pp. 52-53.

5. "African Music," *Encyclopedia Britannica*, 2007. Online 11, URL: http.www. Britannica.com/eb/article. 57074.

6. *ibid.,*

7. J. D. Williams, "The Art of Music," South African Archeological Society Newsletter, 1981, p. 8.

8. *Encyclopedia, op. cite,* 57074.

9. There has been a five year project located within the Free University of Berlin which has been investigating the parallel history of instruments in both art and science. My contribution to this project has been published as, Don Ihde, "Die Kunst kommt er Wissenschaft zuvor. Oder: Provozierte de Camera Obscura de Entwicklung der modernen Wissenschaft?" in *Instrumente in Kunst und Wissenschaft,* eds. H. Schramm, I. Schwartre, J. Lazardzig (Berlin: Walter de Gruyer, 2006), pp. 417-430.

10. Still following a long interest in instrumentation, a number of my works are now beginning to be published, see Don Ihde, "Models, Models Everywhere" in *Simulation: Pragmatic Construction of Reality, Sociology of the Sciences Yearbook 25,*eds. J. Lenhaard, G. Kuppers, T. Schinn (Dordrecht: Springer 2006), pp. 79-86.

11. Trevor Pinch and Karin Bijsterveld, "Breaches and Boundaries in the Reception of New Technologies in Music", Technology and Culture. 44.3 (2003), p. 538.

12. *ibid.,* p. 540.

13. Martin Heidegger, "On the Hand," cited in Michael Heim, *Electric Languagae: A Philosophical Study of Word Processing* (New Haven: Yale University Press, 1999), pp. 210-211.

14. Felix Hess, *Light as Air* (Heidelberg: Kehrer Verlag, 2001).

# 4

# Phenomenologists and Robots

Imagine a room in which there might be an equal number of somewhat humanoid *robots* and a bunch of *phenomenologists*. One might be forgiven for thinking such a pairing to be—at the least—odd. This would be particularly the case if the phenomenologists were 'classical' phenomenologists who would be thought to be philosophers of consciousness, doing analyses of 'subjective' phenomena, but here faced with a bunch of motile, seemingly animated machines—robots. What one would likely expect from such an encounter might be a kind of analysis which *contrasts* the robots with any humans-with-consciousnesses. In this case, the phenomenologists simply deny that the robots have consciousness, experience, or perceptual capacities of any kind. At most these machines might 'mimic' to some clumsy degree, what 'appear' to be human-like gestures or sounds or whatever. The robots, these phenomenologists might hold, are simply more animated versions of Cartesian cleverly designed automatons.

Historically, or outside this thought experiment, phenomenologists have actually been interested in robots, automated machines and the like—at least since mid-century last and under the primary influence of the work of Hubert Dreyfus. Of course Dreyfus did not address robotics in his early work, but rather he became the critic *par excellance* of early forms of *artificial intelligence*.[1] It was later that robotics became interesting, but as an indirect result of the impact of the Dreyfus critique. I cannot here go into great depth concerning this history, but want to suggest a few important outcomes from this mid-$20^{th}$ century controversy.

First, cast in lines suggested by the thought experiment, one strand of those hoping to create artificially intelligent machines— usually with computers rather than robots—would have been happy with the contrast and mimic rhetoric. This strand of interpreters would argue that machines *can indeed be intelligent, artificially intelligent* and employing the usual 'inevitable progress' myth common to modernist narratives, would claim that it is only a matter of time and improved technology until this happens. Here one finds the Marvin Minskys, Hans Moravecs and Ray Kurtzweils. This group of thinkers include doing fantasies of 'downloading human intelligence into a computer.' There is, however, a reverse side to this strand. The reverse side could be called the 'reductionist' strand which actually implies that human and machinic intelligences or experiences are ultimately identical—but with the identity now reducible to the *machinic* or mechanical. This, too, is of course modernist and continues to echo Descartes in one of his phases.

Dreyfus, as we know, strongly contested this two-sided strand of interpretation and while he, himself, remained modernist insofar as he could still be called a philosopher of consciousness, gave the controversy a major twist. This twist was the interjection of a Merleau-Pontyean emphasis upon *embodiment*. For Dreyfus, all intelligence presupposes a *body,* and for practical purposes a *human body*. And, indeed, one of his primary arguments against artificial intelligence took the shape of the claim that a computer cannot 'think' or be intelligent *because it has no body!* Now one of the secondary implications embedded in this controversy which led to a phenomenological interest in robotics, was the implication that perhaps it would be *harder* to model motility machinically, than intelligence, particularly of the calculative kind. Could those who follow the mimic trajectory, successfully produce a *graceful and agile* robotic motion?

There is a second aspect of the Merleau-Ponty-Dreyfus move to bodily motility which holds a deeper problem for our phenomenologists and robots. And that is what I shall call the *move away* from phenomenology as a philosophy of consciousness which arises from the emergence of embodiment. Examples include the activities of riding a bicycle or typing on a keyboard. In both cases 'consciousness'—at least in its Cartesian central control model—is transformed and displaced. If I try to tell myself "turn into the direction of the fall" of the bicycle, I am more likely to fall than not. Whereas if I have begun to 'embody' the bicycle into my now extended bodily awareness, I instead focus upon the pathway I am following and remain in balanced motion. Or, similarly, if I ask myself which key do I now type in order to spell "phenomenology," I am either going to type too slowly or disrupt the flow of typing altogether, whereas if I simply see the word taking shape upon the screen without any focal awareness of where I am placing my fingers, all goes well. The embodiment activity thus *transforms* and in some sense *displaces* consciousness.

## 4. Phenomenologists and Robots   39

In Dreyfus's development of this modification of consciousness, the result is his complex and well known phenomenology of expert action or expertise.[2] According to Dreyfus, one starts with what I shall call 'executive' consciousness as displayed in rule-governed behavior. "To shift gears, one first depresses the clutch pedal while moving the shifter to first gear...." But then, with practice, this motion becomes 'automatic' and no longer needs any executive conscious activity, the motion is 'embodied'. The skilled actor transcends executive consciousness as in rule behavior, and instead plays out *patterned* behavior seemingly without explicit consciousness. This, however, carries possible serious consequences for phenomenology! A first and minimalist pass at the implications for consciousness, would be to recognize that consciousness does not disappear, the expert typist, bicyclist, or car driver does not become *unconscious,* although if Dreyfus is right, some aspects of previously explicit consciousness do disappear or almost disappear from awareness. In the typing case, I cease to be aware of what keys are where—in my own experience, if you were to give me a picture of a keyboard with no numbers or letters on the keys, I would find it hard to correctly fill in the blanks, although if I had to type this paper on a keyboard without its letters and numbers on the keys, I would have no trouble operating it. My 'consciousness' when engaged in typing, is *not focused upon the keys,* but *through the keys* and perceptually focused upon the text which is being produced on the screen. In what I call an *embodiment relation,* I have embodied or taken the keyboard *into* my now extended bodily experience.[3] I am very little aware, or am at best *fringe* aware of the keyboard, although I still do feel my fingers doing the typing. More descriptive would be the Merleau-Ponty notion that "I am outside myself in the world."

To summarize what has changed: 'consciousness,' now in its Merleau-Pontean-Dreyfus sense, has drastically changed. It is no longer a 'subject in a body,' nor is it a central, executive consciousness. It is, instead, an embodied, motile consciousness not identical with a 'skinbag' body as Andy Clark puts it.[4] It is directionally 'outside' such an enclosed body, in an experienced world. To my mind these are moves in the right direction, but also, I will hold, they are not yet radical enough to overcome the problems to which I pointed in the thought experiment. In that experiment, the phenomenologists and the robots remained, as it were, two distinct kinds, one conscious and subjective, the others machinic and object-like. Were this the case, there would remain a sort

## 4. Phenomenologists and Robots

of ghostly echo of Cartesianism in the experiment, in which case the phenomenologists are the mind-surrogates and the robots the body-surrogates, with the dualism now the division between the humans and non-humans. To exorcise this ghostly echo, I shall return to Dreyfus with some attention to the areas where he and I might disagree, and from this I shall begin my itinerary towards *postphenomenology*.

I begin with a little personal history related to Bert Dreyfus and me: We originally met at MIT in 1958 where we both worked until I finished my Ph.D. in 1964 when I moved to Southern Illinois University, and he, some years later, to Berkeley. Our paths often crossed and we frequently were on the same programs, particularly those involving computers, AI, and other technology-centered conferences. One of the things I found amusing was that whenever Dreyfus claimed that AI or computers **could not** do something, he unintentionally helped set up the next years of programming for the computer jocks—they would inevitably attempt to do programs which would prove computers **could** do what Dreyfus claimed they could not. I shall later look at one of these programs, the Big Blue chess program, which the computer jocks claim totally defeated Dreyfus's claim, but which may be more complicated than appears in the literature around Big Blue's defeat of Gary Kasperov. I did think, from the beginning, that chess playing programs might well lead to machinic competences which would equal or exceed human ones. I had three reasons: the first was that chess, after all, is a rule-governed 'world' which is quite unlike the open textured lifeworld, or for that matter, the actions required to speak and understand a natural language. The second reason relates to notions of *bodies*—I disagreed with Dreyfus that computers do not have bodies. Clearly they do not have human or for that matter, animal, bodies, but they have electronic 'bodies' which materialize their actions, albeit in different ways from those of living, organic beings. The third reason, which I think remains deeply phenomenological, is that when it comes to humans, computers, or robots, the relationships are *interactional,* or, *relativistic.* This is a phenomenological ontology which focuses upon such inter-relationalities. And it was this last reason why, from the early days of Alchemy and Artificial Intelligence, Dreyfus's 1967 report to the Rand Corporation, that I felt uneasy with the way he framed the critique. And the same applies to his later highly successful *What Computers Can't Do,* through its various revisions. Dreyfus says a lot about how computers and purported artificial intelligence are *different* from human intelligence, but actually very little about the importance of the human-technology *interface* or of the often hidden implications therein. It is at this juncture that I shall now begin a positive analysis which will simultaneously emphasize the way a *postphenomenology* begins to revise and modify classical phenomenology, and to—by way of concrete examples or case studies—show how a reframing of the inter-relational role of our 'phenomenologists' and 'robots' takes quite a different direction from the older Dreyfus frame. I begin with *embodiment:* There is a broad sense, in my and what I call a postphenomenological context, that embodiment effectively replaces the classical phenomenological sense of *subjectivity*. As a growing literature recognizes – *Chasing Technoscience* (2003) and *Postphenomenology: A Critical Companion to Ihde* (2006)—both claim that postphenomenology is a non-subjective phenomenology.[5]

That may seem, at first, strange. Embodiment continues to be *experiential,* and in a preliminary sense it also begins with first person experience. But embodiment, as with the Merleau-Ponty-Dreyfus moves, recognizes the focal role *being-a-body* has for all human experience, activity and knowledge. Nor are we humans just any body—we are a particularly structured body which shapes our actions, experience and knowledge in quite particular and also in constrained ways. Our action, experience and knowledge is *situated*. I shall use some perceptual examples both to show shapes of situatedness and to reveal how our learned awareness of these shapes is attained *interactionally*. Take vision: we are normally binocular, forward oriented in our vision. In several recent publications, Donna Haraway and I have exchanged work on this forward orientation.[6] First, if one *varies* in good phenomenological fashion, other shapes of vision, one can indirectly gain greater insight into the particularity of human vision: raptors, hawks, cats and dogs, apes, share our forward vision. But rabbits, whales, tortoises and other creatures who have eyes on the sides of their heads do not. And, if we do a proper analysis of how we learn this self-knowledge, something else emerges. I claim *we do not have direct, introspective knowledge of our visual shaping*. Rather, this self-knowledge must be gained *reflexively* and in strict *interaction* with our experience of being-in-a-world.

It might be the case that we learn to be aware of our forward oriented vision and the role of our eyes, by seeing ourselves in a mirror with the self-recognition capacity we also have, but which we also share with elephants, dolphins and some apes. Or, perhaps seeing other humans and recognizing their shapes, we refer back to ourselves with the *other-as-mirror*. Or, if we are really good phenomenologists, we might take account of our experiencing of objects which, following Husserl, always present themselves in *profiles, adumbrations* such that we can either get a more adequate sense of their three-dimensionality by walking around the object and thus gaining multiple profiles, or having the object rotate itself for us to have another way of gaining multiple profiles. Note that this dialectic of stability/motion—either the object is stable and I move, or I remain stable and the object moves—is also repeated at the micro-level with vision. If I stare at some object with a fixed gaze, the object will soon appear to 'wobble,' and cease its stability; but to keep the object stable, I must ever so slightly move my visual focus across the object, thus engaging micro-bodily motion. In all these cases there is an inter-relational situation in which one moves progressively towards greater adequacy by means of motile, bodily variations in coordination with other beings in the world. In all of this we learn about our directional, shaped and situated structures of vision—but this is not *introspective*. It is rather, *inter-relational* and *reflexive* and in each case involves 'external' perception.It is only in relation to our surrounding world and our interactions within it that we learn of the shapes of our visual situatedness. Whatever knowledge we so acquire is *inter*-subjective and *inter-objective*. Such inter-relationality, I claim, is not 'subjective' but relativistically is a rough organism-environment self-knowledge attainment. (And while I shall not follow this out here, this is one of the reasons I have grown to appreciate John Dewey's pragmatism as a first layer of critique of early modern epistemology, over Husserl's still too 'consciouness' laden approach to this phenomenon. Both Dewey and Husserl began their respective experiential based philosophies with attempts to deconstruct early modern epistemology—Husserl's target was often Descartes; Dewey's was Locke. But both Descartes and Locke held what I call a *camera obscura* version of subjectivity-consciousness, namely the 'subject' is located inside the camera-body and experiences only the images, sensations, etc. which come from the 'outside'. This mind-in-a-box is quite contrary to the Merleau-Pontyan 'outside myself in the world' and an inter-relational embodied self.) With these variational examples I am trying to show how *embodiment* replaces *subjectivity* –at least the 'mind' subjectivity of early modern epistemology—in a now modified *post*phenomenology. In addition I have shown how *intentionality*, the classical name for the reflexive, interactive shape of experience plays its role as well. It is now time to re-insert the machinic into this discussion. Perhaps the best known hallmark of my earlier work in the philosophy of technology, is my "phenomenology of technics" which first was published in *Technics and Praxis* (1979), and expanded and refined in *Technology and the Lifeworld* (1990), and reprinted in virtually every major collection book on philosophy of technology. I shall not here repeat that analysis, but only point out that from the early 70's on I held to an interactive ontology which would make the smallest 'unit' of analysis the human designer,user, or other action types *plus* the technology being designed, used or acted with. In older classical terms, this was intentionality with, through, alongside the materiality of technologies.

Embodiment, now as phenomenologically described actions, was also there from the first. *Embodiment Relations*, I claimed, were those human actions *through* technologies directed at some effect in our environing world. In this style of inter-relation, my eyeglasses 'withdraw' as objects, become 'transparent' in letting me see the world, or, if riding my bicycle, I embody it and through it experience the roadway and its surroundings with the balanced speed attainable with 'wheels.' Or, by developing particular skills in focused ways, *Hermeneutic Relations* based upon our interpretive abilities, could be used to 'read' instrument panels, gauges, and the like, which also refer to the world in some of its dimensions. Or, with *Alterity Relations* one can inter-relate *with* a technology, in this case including our robots, as quasi-others and the interaction focused upon the technological object. In all these relations, there remain specific roles for embodiment and particularly shaped interactions of humans vis-à-vis technologies.

I shall now turn to a series of concrete variations upon human and machinic phenomena and show how a postphenomenology is also 'postmodern' in some sense:

- One of the background features of modernity—particularly in the West or Eurocentric contexts—is the role of *human exceptionalism*. Human exceptionalism lies deep in our religious history, both Jewish and Christian and continuing through Islam. It is the belief that humans are the epitome of creation; superior to animals—and by extension, machines—and have religious rights to 'have dominion' over the rest of creation.

## 4. Phenomenologists and Robots

- Postmodernity, in the late $20^{th}$ and early $21^{st}$ centuries, has begun to call human exceptionalism into question. This is most dramatically the case with human-animal critical studies. Much recent literature has indicated two 'paradigm shifts' which have taken hold with animal studies: first, there has been an opening to an 'animal theory of mind.' Now, while I think this phrasing is actually archaic with its implicit reference to the clearly Cartesian, analytic notions of mind, its effect is more radically postmodern. Animal ethology has now rejected both the past mechanistic and behaviorist reductionism related to animal intelligence, and animals are now studied with respect to their styles of intelligences. As it now turns out, ravens, crows and jays may frequently be better at calculating and solving puzzles than many primates! Second, animal studies have now shown how various animals have 'cultures'—practices which are learned and passed on from generation to generation, many of which include *technologies and tools*. Contemporary animal ethologists have classified approximately 27 distinct Chimpanzee cultures. I earlier referred to the self-recognition animals: humans, some primates, dolphins and elephants (so far), and each of these animals also use tools. Chimpanzees fashion probes to extract termites, use clubs and hammer/anvil tools; elephants use large tree pieces for various activities; and most surprising, female dolphins use sponges to protect their beaks while getting shellfish from ocean bottoms—a recent discovery of such a dolphin culture in Australia. Both these paradigm shifts, which allow animals minds and cultures, tend to situate animal actions closer to us humans, rather than farther as in modernity.

- For purposes here, I shall temporarily avoid the 'language' question which also arises in animal studies, because I want to begin to focus upon machinic questions. Instead, I want to make some much simpler comparisons about relative 'superiorities' with humans, animals and machines as illustrative. And in keeping with the embodiment emphasis, I turn to bodily capacities in these comparisons. I begin with speed. Surely no one can hold that a human runner has superior speed compared to either animals or machines! The fastest human runner attained 12 m/s in a short dash; a cheetah easily attains 30 m/s in a sprint, and even a hippopotamus can outrun a human. The contrast, when compared with any

## 4. Phenomenologists and Robots    45

range of technologies is even much more stark: a speeding bullet, an unmanned jet drone, or any number of devices far exceed our bodily capacities for speed. The same comparisons could be done with visual physiologies: human vision, once diurnal, is normally color-sensitive, discerns depth, etc. but remains considerably short of bird vision which has a wider color range (into ultraviolet and infrared in many cases), and a much finer figure/ground discrimination capacity, and on and on and on.

- Now the machinic: Human-technology variations turn out to be exceedingly tricky, much more so than human-animal variations in my estimation, because the *human* in human-technology relations is so deeply implicated "all the way down." Now in 'origin myths' or even more ordinary contexts, *found objects* can be turned into 'technologies.' Thus our most ancient ancestors could have picked up rocks to use as thrown projectiles, or sticks to use as clubs (or used similarly by Chimpanzees). But a second step is to not only transform a found object, but to *materially modify* the object to fit its use-context. Thus the more than million year old Acheulean hand axe, which modifies stones, becomes a pre-*homo sapiens sapiens* technology. If we skip now to the present, in this trajectory relating to human and technology capacities, imagine now, video games: Many such games are designed to have different 'levels' of play. So many are variations upon enemies attacking—at each level more and more come, until it becomes *impossible* for the human player to fend them off, or, in another variation, the enemies come faster and faster, again making it impossible for the human player to react quickly enough. This excess of technology speed or numbers also belongs to a wide range of other technologies as well. My son once had a composition program in which a digital piano was connected to a computer—he could play whatever he wanted and the program would print out a score which he could then edit. But, he could also adjust things so that instead of a ten finger, one-player constraint, he could program in as many fingers as he liked—20, 30,40— or, alternatively, jack up the speed such that no human could play that fast. All of this is again human-technology interrelational.

- The examples I have just given, one can imagine, might be

## 4. Phenomenologists and Robots

taken by a disappointed modernist as *dystopian*. Because our very machines now 'exceed us' does this not show how 'technology' undercuts our very *humanity*? This theme became that of so many of the first generation philosophers on technology, echoed by Karl Jaspers, Martin Heidegger, Hans Jonas, et.al. There is, however, an unexpected other side to the human-technology interaction described above. Playing with the types of technologies—video games, composition programs, and the like—also have led to honed new skills for the users. The mere "eye-hand" reduction bemoaned by some, has become in a new setting what is today often called "Nintendo surgery." Laproscopy, robotic prostate surgery, angiograms, surgeries performed with tubes equipt with optical, surgical tools, all miniaturized and inserted with minimum skin openings, call precisely for the similar eye-hand plus visualization on screens skills. I, myself, have experienced this as a patient in two angiogram processes and unconsciously with heart surgery. Note that while highly specialized and engaging only certain bodily skills, all this is clearly related to my notion of *embodiment relations*, where the skilled surgeon is able to 'embody' through his hi-tech instruments, the required action within the patient.

- If we switch the variation, now to distant sensing, for example the various Mars robotic vehicles, Sojurner Truth, the Mars Explorer, and the like, one must modify the immediacy of Nintendo surgery, for the more complex and more hermeneutic relations of programmed commands. Transmissions to and from Mars entail time delays, and many of the instructions or commands to the vehicles are 'typed in' rather than manually operated. But the human-technology relationality remains central to this mode of human-robot context.

These variations are the first step in my postphenomenological re-framing of the human-technology context. By focusing upon embodiment and inter-relationality, I have shifted away from the separation of the machinic and the human to the interaction of the machinic and the human. Instead of a metaphysical dualism of different 'kinds', by looking at the interactivity of humans and machines, a different picture emerges. I return now to the Big Blue incident and apply this re-framed analysis to this event. Big Blue, the chess playing program, was of course related to humans at all stages. The program did not create itself, but was designed, re-designed and tweaked by humans all through its history. Even Dreyfus lay in the shadows since some of the programmers deliberately wanted to prove him wrong. They studied chess strategies; they developed the sheer massive calculation strategy which operated at superhuman speeds; they worked on the heuristics. Big Blue was from the beginning filled with programmers, as Bruno Latour would put it. But, the managers of the event—the IBM corporation in particular—also had interests in the game. And thus the public relations frame was to cast the event in precisely the 'two kinds' interpretation—'man' versus 'machine'. Here was an echo of a whole series of myths, including the American myth of John Henry who died beating a steam tunnel digger. Only here it was human 'brain' versus the artificially intelligent Big Blue.

Cast in this light, the 'defeat' of Kasperov was touted to be the victory of a chess playing computer over Master Player—but that is not what happened. Instead, after each game, behind the scenes, the small army of programmers continued to tweak and improve the program, learning from Kasperov's earlier moves, and thus draws, and eventually wins, by Big Blue began to occur until Kasperov was 'defeated.' But, I contend, *this was not a 'man' versus machine event at all!* It was man, Kasperov, versus a whole collective of humans plus programmed computer. The interaction was asymmetric and a 'gang' of humans plus computer 'beat' Kasperov. By itself, Big Blue could do very little, and only through a long "dance of agency" as Andrew Pickering would put it, did the honing which finally worked emerge. Stripped of the 'man' versus machine mythology, the Big Blue/Kasperov event may be seen as a very complex phenomenon of human-technology interactions.

I want now to take one more step. I have titled this paper, "Human and Machinic Embodiments," but have not yet directly addressed machinic embodiments. I echo my earlier disagreement with Dreyfus, computers *do* have bodies, or a kind of materiality, albeit different than humans. If this is so, what kind of 'bodies' are involved and how do we learn what such 'bodies' can perform? First, I want to make clear that this part of the analysis will not take any modernist 'physicalist' or objectivist approach to this question: I am not going to talk about electronics, hardware arrangements, chips or the like. Rather, I shall remain postphenomenological and look at machinic embodiment in terms of interactivity and the mutual learning and transformations which occur therein. I am going to take my examples from long term research program on imaging technologies. Several years ago, my colleague, Bob Crease, published a book, *The Prism and the Pendulum: The Ten most Beautiful Experiments in Science* (2003), based upon a poll of physicists choices. It turned out that of the ten, four involved experiments with light or light-like phenomena. In my own research, I had already noted that, beginning with the Renaissance *camera obscura*, a progression of science's imaging machines were variations upon that camera. I published several versions of this early work under an article titled, "Did the *camera obscura* invent Modern Science?"[7] I shall here draw from a piece of that history, a postphenomenological analysis of the human-technology inter-relations which eventuated in whole classes of imaging technologies important to science.

The optical effects of the *camera obscura* were already known in antiquity—Mo Ti in China (400 BCE), Aristotle (350 BCE) and then fully by Al Hazen on optics (1036 CE). It later became a favored optical toy in the Italian Renaissance ($15^{th}$ century). For our purposes the camera has a *light source* (sun or artificial light), an *aperture* (at first a round hole), and a *screen* upon which an image is cast. Early *cameras* were objects of fascination by producing what I call *isomorphic* or picture-like images. DaVinci used one to produce an image of a crucifix in his room, and later Galileo used a version of the camera as a helioscope to image sunspots (1609). But I shall begin later, with Isaac Newton's transformative variation on the *camera* (1666),to produce a *non-isomorphic* image which does not look like the image source. What Newton did was to set up his *camera* by cutting a round hole in his window shades, through which the sun could shine, but then he placed a *prism* at the aperture and the 'white' light which streamed through the prism produced a 'rainbow spectrum' on the blank wall opposite. This was the imaging technology which stimulated his theory of color, including the recognition that 'white light' was a composite of the 'rainbow' of colors—and he ingeniously showed this to be the case by reversing the 'rainbow' back into 'white' light by adding a second prism!

With the wisdom of retrospection, we can note that Newton did, indeed, produce a revolutionary instrument by simply placing a prism in the aperture. But, he retained the traditional 'round hole' aperture which normally produced an isomorphic image. The addition of a prism, however, could only produce a somewhat 'blended' rainbow. Separations between colors were not dis-

## 4. Phenomenologists and Robots 49

tinct, although they were recognizable. Historically, scientists remained fascinated by the *camera,* and here I turn to its non-isomorphic trajectory. Continued experimentation with variations on the prism, did lead Wollaston (1802) and Fraunhofer (1814) to produce much more distinct spectra, but a practical *spectroscope* was not perfected until Kirchhoff (1859) produced one. In this history, however, the crucial change again related to a variation on the aperture—spectroscopes use a *slit* instead of a round hole or lens, and the resultant change in the image entails *clearly separated lines of distinct colors, plus the emergence of 'emission' and 'absorption' lines.* Skipping to the punch line, $19^{th}$ century scientists finally recognized that each part of the now distinct spectra were *chemical signatures,* with one being sodium found being emitted from the Sun. So, now we have two important variants upon the non-isomorphic spectral imaging, Newton's rough 'blended' spectrum to Kirchhoff's distinctly lined spectrum.

This trajectory, however, does not end here. Thomas Young cognizant of earlier history of light experimentation, knew that light through a peculiar crystal, Icelandic spar, produced a *double refraction* with two rays producing different effects. Young returned to a new variation on the *camera* to produce this effect. This time the aperture was given a *double slit.* The image produced was that of intersecting black and white bands analogous to waves *interfering* with each other. The double slit *camera* invents *interferometry* which established the wave structure of light.

For practical purposes here, I end my brief history with these 'origin trajectories,' all variations upon the ancient *camera obscura.* Were we to bring this up to date with contemporary hi-tech versions of these instruments, we would find that whole classes of 'beam splitters,' mirror-plus-optics, and above all 'beam active' light analog sources such as with the coherent light sources of lasers, photon beams, electron beams, ion beams, all producing phenomena at previoiusly unknown micro-levels. Today's spectroscopy, in mass spectroscopy can analyse chemicals at the atomic level, and interferometry, with photon beams can show quantum phenomena, each variation contains new levels of imaging.

While I shall now turn to a postphenomenological analysis of this history, please be aware that my foreshortened and oversimplified history could be deceptive in the sense that the discoveries which have emerged out of these variations on imaging technologies have been hard won and have occurred over vast stretches of time: From sunspots with Galileo (1609) to Newton's theory

## 4. Phenomenologists and Robots

of light (1666) to the chemical compositions of the sun and stars (mid $19^{th}$ century), and from the wave theory of light (1804) it its modern quantum variants ($20^{th}$ century), we stand amidst many of the important attainments of modern science. We also are looking at this history from a perspective which is starkly different from much standard philosophy of science. What we are seeing here is the emergence of discoveries from the very core of human-technology relations. Galileo did not expect to discover sun spots; nor did Newton expect to discover the composite structure of light; and the eventual discovery that spectra were chemical signatures was hard won over decades of time. In short, what this perspective shows, does not map well upon a hypothetical-deductive predictive practice notion of science. It rather shows a much more pragmatic, experimental, and materially based science whose 'observations' and thus discoveries are *technologically mediated* by the instruments of science. Such a 'material' science is fully embedded, I argue, in the *lifeworld*. An instrumentally mediated science is a human-technology-world interactive science.

A more detailed, close-up analysis would show that each of these origin trajectories were attained through human-technology interrelations in which whatever human aims were being sought, an interactive and learning process was engaged in such that the human gradually 'learns' what the material forms of machinic 'embodiment' allow. Bob Crease, for example, told me that when writing about the Young double-slit *camera,* he tried to construct one following Young's description. He succeeded, but only after much tinkering and much effort since the tiny split in the aperture required him to learn considerable skills. Much simpler, of course, are the pinhole cameras one can construct to view eclipses—a number of which I, myself, have made and successfully used. One of the best case studies on science-instrument interactivity can be found in Andrew Pickering's *The Mangle of Practice* ( 1995)[8] He describes well the 'resistance and accomodation' discovered in instrument development, the 'dance of agency' between humans and machines, the 'tuning' which is necessary, and so on. But, once the interactive skills and tuned instruments begin to work in human-technology conjunction, then an origin trajectory can lead to greater and greater refinements, variations, and progressive development as my suggested history has shown.

Note how, in this postphenomenological, interactive style of analysis, that the skills learned in developing and using instruments parallels quite precisely the same interactive pattern noted

## 4. Phenomenologists and Robots 51

in becoming bodily self-aware. One neither is *directly* aware of the possibilities or constraints of the instrument, nor does one *derive* this set of capacities from simple material properties. Rather, it is in use that one reflexively becomes aware of such capacities. The same phenomenon can be seen in musical examples: performers must always *learn* and usually through extensive practice sessions, the capacities of their individual instruments. I have been suggesting that in this human-technology interactivity, the skills learned to use the instrument, and the reflexively gained awareness of possibilities and constraints, emerge in this active, motile exchange.

However, very often in the history of such interactivity, *play* is also a factor in the acquisition of our awareness of machinic embodiment. I will give a short, autobiographical example: A few years ago I was brought to a technical university in Umea, Sweden, as the external examiner of a Ph.D. candidate who had previously been a Visiting Scholar in my Stony Brook group. I was invited to enter and experience a 'virtual reality cave' which his laboratory had set up. I was given a pair of 3-d goggles, and hand set to control motion and a few instructions as to the purpose of the 'game' involved. The scene was that of an island, in which there were buried caverns and the object was to find a magic key which would allow me to get treasure or some secret. At first I felt rather clumsy as I accommodated to the strange new quasi-bodily motion which remained quite unlike that of normal motility. This was much more a sort of visual, 'floating' style of movement than the usual upright posture, forward walking motion. But, soon, I discovered that I could 'fly' virtually and as I became more accommodated to this 'virtual' motility, began to experiment, or *play* with the new motion possibilities. To cut the story short, I soon decided to 'escape' the island and 'fly' off the cliff which formed the horizon of the scene, so accelerating as fast as my hand set allowed, I headed for the cliff—only to have the whole scene abruptly halt. The program did not allow such an escape, or one could say, I discovered a constraint-limit in this virtual 'reality' by means of the play I had undertaken. Again, in this case the quasi-world of the virtual scene reveals its embodiment through my failed interaction.

I return now, one more time, to the history of spectroscopy and the hard won recognition that spectra reveal chemical signatures. Spectra with sharper color distinctions than Newton's 'rainbow' began to be recognized early in the $19^{th}$ century by Wollaston and Fraunhofer. By 1849, Leon Foucault noticed the similarity of a part of the yellow spectrum from a sodium lamp matched part of the Sun's spectrum, but did not recognize the reason for this. Finally, Kirchoff who perfected a practical spectroscope, realized that such signatures, whether on earth or solar, were those of the same chemical source. And, once so recognized, by compounding the technology of the telescope with a spectroscope, star analysis got off the ground and began to produce new classifications of stars according to their different compositions. Admittedly, the questions raised in science are different from those raised in other human activities, but the human-technology interactive learning process has the same phenomenological features. Musical performance, another human-instrument practice, produces its new sounds from the same playful, experimental, human-machinic interactions. So, we end with not two 'kinds', phenomenologists and robots, but with a *dance* of human and machinic bodies, in performances of many kinds with as yet unforeseen results.

**Endnotes:**

1. Hubert L. Dreyfus, *What Computers Still Can't Do* (Cambridge: MIT Press, Third Printing, 1993). Dreyfus, often in collaboration with his brother, Stuart, has long been a gadfly to the mid-20$^{th}$ century school of utopian artificial intelligence school. Beginning with a RAND Corporation study in 1967 and then to the first edition of *What Computers Can't Do* (1979) there has been a long standing debate concerning human and machinic thinking within which Dreyfus has played a major critical role.

2. Another of Dreyfus's major works has centered on an embodied, phenomenological theory of learning and expertise may be found in Hubert Dreyfus and Stuart Dreyfus, *Mind Over Machine* (New York: The Free Press, 1986) and also in the context of a debate in Robert Crease and Evan Selinger, *The Philosophy of Expertise* (New York: Columbia University Press, 2006).

3. Don Ihde, *Technics and Praxis* (Dordrecht: Reidel Publishers, 1979) and then *Technology and the Lifeworld* (Bloomington: Indiana University Press, 1990) develop a phenomenology of technics, analyzing a continuum of human-technology relations.

4. Phenomenologists and Robots    53

4. Many cognitive science, neurology and philosophy of mind thinkers have begun to respond to notions concerning the plasticity of the human body, including Andy Clark in his *Natural Born Cyborgs* (Oxford: Oxford University Press, 2003).

5. In both Don Ihde and Evan Selinger (eds) *Chasing Technoscience* (Bloomington: Indiana University Press, 2003) and Evan Selinger (ed) *Postphenenology: A Critical Companion to Ihde* (Albany: SUNY University Press, 2006), a set of authors from a wide variety of disciplines discuss the development and role of a postphenomenological analysis. See also the special issue of Human Studies, 31/1, 2008, for examples of postphenomenological research.

6. Donna Haraway and I have exchanged articles on human and animal vision in counterpart volumes. Hers, "Crittercam: Compounding Eyes in NatureCulture" first appears in the Selinger *Postphenomenology: A Critical Companion to Ihde* (2006) volume, pp. 175-188 and mine, "Predatory Vision" is forthcoming in a volume, *Thinking with Haraway*.

7. Don Ihde, "Art Precedes Science, or, Did the *Camera Obscura* Invent Modern Science?" in *Mediated Vision*, ed. Petran Kockelkoren (Arnhem: ArtEZ Press, 2008), pp. 24-37

8. Andrew Pickering, *The Mangle of Practice* (Chicago: University of Chicago Press, 1995)

# 5

# Beyond Embodiment and a Return

This last September at an ancient Buddhist Temple in Kyoto I had my first opportunity to experience "sound beyond sound," an experience known centuries ago to monks of this tradition. Large brass gongs are placed within this temple, and one may pick up the striker and make the gong "sound," which it does for a long time, although eventually the sound fades out and one can no longer hear it—*but, if one places one's hand on the gong, one can continue to feel the vibrations for a significant time past one's ability to hear the gong.* This is "sound beyond sound," a phenomenological recognition that there can be sound beyond human hearing.

Today, in our presumably sophisticated scientific world, we recognize that sound beyond our human experience of sound reaches from *infrasound,* sound below the horizons of our hearing, to *ultrasound,* sound well above our horizon of hearing. And in bioacoustics we have had recordings of 'whale songs' from infrasound ranges for some decades, but only more recently 'mouse songs' from ultrasounds have been added to the echolocation sounds of bats longer known. Many of our animal relations can hear what we cannot, sound beyond sound at both lower and upper limits. Indeed, I use ultrasound devices in my Vermont house to deter mice, bats and other creatures which might like to winter invade my house.

I begin with these simple and by now familiar examples which in their very mundanity mask the complexity of technoscientific implications for human embodiment and technoscientific instrumental mediation. It is that complexity which I want to unpack on this occasion. I will begin with *embodiment* in perhaps its most immediate and experiential sense—so long as I am aware, conscious, I continuously *perceive multidimensionally,* I see, I hear, I feel, and so on. And while I may switch my attention from one dimension to another, what remains continuous is a *whole-body experience* of my immediate environment. I can selectively focus upon sight or listening, but I cannot turn them off. This is, simply put, old fashioned *phenomenology.* I can be aware of my experiencing of a world. And while I could go on in great detail about the shape of this experience with its figure/ground, attention shifting, and other configurations, for purposes here I want to focus upon what phenomenologically would be called *horizonal* experiences—how do I discover my 'limits'?

I return to auditory experience: by the time I had reached my sixties, I had a vague awareness that my hearing was changing. It had become harder and harder to focus upon conversations at parties or in large groups. Background sounds seemed to overwhelm my ability to pick out only the relevant conversations. Then, one evening with a seminar in my home, we were listening to a CD of forest sounds of water dripping and running from Steven Feld's example of New Guinean forest dwellers and all the students and my faculty colleague, obviously were responding to these sounds—and I recognized that I *was not hearing* what they were hearing! That same year, in the science museum in Boston, my son had me listen to a frequency-producing machine which one could tune from lower to higher frequencies which when I operated this machine I discovered that I could only hear up to about 11,000 cps, well below the 20,000 cps I knew expert listeners in their prime could hear! I was discovering my auditory horizon, my hearing limits. But—how was that possible? Experientially, I could listen up to a limit which while the sound gradually faded out, once one dialed no matter how high, there was no experienced sound—I could not experience beyond my limit. Using this experience of reaching a limit as a *diagnostic,* I now want to turn back to sounds beyond sounds, sights beyond sights in a new variation.

There is a conundrum located here: clearly I cannot (directly) experience beyond the limits of my experiential horizons, nor can I even place my embodied, situated experience there perceptually— I cannot recognize my limits by getting outside myself and my situation, noting it from 'above' as it were. Yet, I recognize that I have come up against my perceptual limits—but how? My answer is *by means of technological, instrumental mediation.* Indeed, in all my examples above some material technology has been involved. But the mediations which make my awareness of limits possible, function in multiple ways.

**Positive and Analog Mediation: The First Scientific Revolution**
Nature magazine in its 1 January 2009 issue celebrates the UN declared "Year of Astronomy" and specifically recognizes the $400^{th}$ anniversary of Galileo's 'perspicillum,' now known as the tele-

scope. In typical science magazine rhetoric the bulk of this celebration is focused upon the new developing and planned telescopes such as the James Webb Space Telescope, the Large Synoptic Survey Telescope, the European Extra Large Telescope and the Square Kilometre Array. But they also include a retrospective by Owen Gingerich on Galileo's perspicillum. He claims that, "what Galileo had done was to turn a popular carnival toy into a scientific instrument." **(1)** And, it is the case that many of the primary scientific instruments of the first scientific revolution were optical—telescopes, microscopes, variations on the *camera obscura*. **(2)**

If I now place this into my frame of reference in which *science is technologically embodied through its instruments* we now see a second sense of *embodiment*. Scientific observations are instrumentally embodied. Gingerich notes that Galileo used his "spyglass to discern craters, mountains and plains on the Moon....by January 2010, Galileo had found four bright moons of the planet Jupiter... [and} also found the phases of Venus..." **(3)** These phenomena—which were to have such dramatic results for the first scientific revolution—had not been previously observed (perceived). But, they were not perceived directly by Galileo either, but through the mediation of an instrument, a technology. These were genuinely new perceptions, and perceptions which changed the world for the perceivers.

I shall not here go into great detail concerning my earlier modifications of classical phenomenology as may be evidenced in both *Technics and Praxis* (1979) and *Technology and the Lifeworld* (1990) in which I analyze telescope observation as an example of *embodiment relations*. I note simply that in these cases, the human experiencer [Galileo] sees the Moon mountains, Jupiter moons, Venus phases *mediated* through the telescope which 'magnifies' his vision in a previously unavailable way. In this context I shall call this an 'analog' mediation since the now first-time seen Moon mountains remain within the limits of humanly visible light in the optical spectrum. Moreover, the Moon mountains are clearly and easily seen as variations upon 'eyeball' visions of the Moon. (For a much more detailed analysis, see the above referenced works). The telescope, like Heidegger's hammer and Merleau-Ponty's blind man's cane, is taken into my now extended and mediated bodily experience:

(Human-instrument) > World phenomenon

To recognize that directed, intentional human experience can *embody* a technology is a first step towards *post*phenomenology

## 5. Beyond Embodiment and a Return

in that a material artifact can be taken into first person experience ( allowing debts to Heidegger and Merleau-Ponty as post-Husserlian). The only more subtle point I want to draw attention to before returning to my current embodiment theme, is to note that the embodiment relation is *inter-relational*. Galileo's Moon was magnified—and that is what he pounced upon—but so was his bodily motion (which he ignored). A hand held telescope magnifies both the Moon and our bodily motion and thus makes it hard to maintain a fixed focus upon our observed heavenly body. Here is yet another clue to the complexity of embodiment: every change in our newly magnified world is also a change in our embodied experience.

I shall call this first technological mediation, which extends already extant visual capacities, a *magnificational* mediation. This same 'analog' capacity pretty much belongs to the whole continuum of early modern optical instruments: telescope (for distant phenomena), the microscope (for micro-phenomena), *camera obscura* variants (for 3d to 2d transformations) and the like. In this step I now have extended direct bodily-perceptual experience in its classical phenomenological sense, to include instrumentally mediated bodily-perceptual phenomena made present through technologies, thus extending the classical phenomenological sense to include material mediational capacities, into an extended sense of embodiment. Were we to return to Galileo, who was clearly aware of how his optics expanded both the macro- and micro-worlds, one could understand why and how he dubbed such mediated vision *superior* to what he took to be limited normal vision. (4)

If we were now to apply Thomas Kuhn's description of scientific practice to this history, we might well call the Galilean development and use of optical technologies a "revolutionary" moment in the history of science, and once adapted as it was from the $17^{th}$ through most of the $19^{th}$ centuries, what followed was a "normal" period of refinement and improvement of this set of optics. For example, Galileo improved his telescopes from an initial 3X to 30X –which accidentally turns out to be the limits of magnification without the *chromatic* distortion which sets in at higher refractive telescope magnifications. Isaac Newton, of course, a century later, *through a new variant on the camera obscura*—this time a darkened room with a prism instead of a lens placed at the aperture—displayed the color spectrum and its separation of colors which he recognized to have what we now call different frequencies. He developed the *reflective telescope,* correcting for

## 5. Beyond Embodiment and a Return    59

this difference by using a parabolic mirror, which overcomes the chromatic distortion and thus opens the trajectory to even higher magnifications. But, all this remains easily understandable within the limits of analog magnification. Note, however, each change, each improvement *entails new instruments, technologies.* (5)

**The Second Scientific Revolution: Beyond Embodiment?**

Return now to our Year of Astronomy and Nature magazine's celebration of the telescope: As I indicated, Nature celebrates the newest developments, but I want to highlight how these now electronic telescopes are radically different from the 'first revolution's' analog magnification. I begin with the Square Kilometer Array. Not yet even sited or built, this will be the biggest *radio telescope* undertaken. Radio telescopy, historically, was the first non-optical process to break out of the limits of the optical range of 'light.' Discovered accidentally from the invention of radio and radar technologies early in the $20^{th}$ century, operators listening to receivers often heard *static*. Then through a classical human-technology learning processes, the operator soon found he/she could discern the direction from which the static came, distance by means of volume, and at first, the immediate cause—thunderstorms. Later, with improvements, sun spot activity could be recognized and much later—leading to a 1978 Nobel Prize for Robert Woodrow Wilson—the recognition of the background radiation of the universe itself. However, this 'acoustic' signaling from afar, was not optical, but in a sense it remained analog insofar as it was a technological mediation of sound-like phenomena. Yet, unlike visual sightings, there had not been any 'earhole' awareness parallel to 'eyeball' awareness of celestial 'sounds' compared to celestial sights. The heavenly spheres, if not musical, were at least static-producing and thus became 'sound sources.' Moreover, celestial radio sources do not always coincide with optical sources—radio astronomy could detect 'dark' or non-optical sources which is precisely what the Square Kilometer Array is designed for: possible detection of 'dark matter' and 'energy.' Nevertheless in phenomenological terms, radio astronomy remains within the analog range of humanly embodied instrumental capacity. It magnifies beyond the ranges of phenomena magnified visually since radio phenomena were experientially more distant.

With the James Webb Space Telescope, we return to the optical spectrum. This planned orbiting telescope is a successor to the Hubble telescope, but with a variant twist. Whereas the Hubble covered the optical spectrum *up into the ultra violet*. The James

# 5. Beyond Embodiment and a Return

Webb takes the optical spectrum *down into the infrared*. In both these cases we reach into a new problem for embodiment. Both ultraviolet and infrared lie beyond the human visual horizons. Thus the mediation provided by both Hubble and James Webb, is different than the analog magnification processes previously noted. I shall call this form of mediation a *translation mediation*. I note several points concerning translation mediation:

- First, while we now reach the detection of phenomena which lie *beyond* the bodily-perceptual horizons of the human, what is detected retains what I call its *instrumental realism*. 'Real' signals are detected by the mediating technology which receives what is coming in. As we shall see, this is the case with $20^{th}$-$21^{st}$ century technologies which now pick up data from nanoscale gamma waves to kilometer scale radio waves—what could be called a 'posthuman' instrumentation which mediates all along the electromagnetic spectrum. And this is part of what I am calling a second scientific revolution, since this instrumentation first appears at the very end of the $19^{th}$ century but gains its primary momentum in the $20^{th}$-$21^{st}$ centuries.

- But, for embodied humans whose observations are those of bodily-perceptual creatures, to become available the information, data, or image must be transformed, translated, into what is open to our *anthropological constant, an embodied human*. Thus embodiment does not disappear, instead it must be taken into account through the new form of technological mediation, translation mediation. This may seem like an indirect acknowledgement of embodiment, but it is a necessary one.

- As noted in Nature, the new telescopes are all 'electronic'—which is to say that computerization, digitalization and all that goes with this $20^{th}$ century development—is implied. And while I have extensively analyzed this process more fully elsewhere **(6)**, here I will emphasize only a few salient features. Telescopy, from gamma wave to radio wave detection, has its information gathering processes linked to *computer tomographic* capacities, and one such feature relevant here is the capacity to transform data into image, and image into data. Thus, if there is a fly-by to radar-map the surface of Venus, the radar probe, like its acoustic analog,

## 5. Beyond Embodiment and a Return 61

can penetrate the cloud cover and map craters in picture-like images. But these must be transmitted 'linearly' and thus the initial imaging is transformed into radio signals for transmission to the home base, and then is re-formed into images for the observing astronomers. Such images *take account* of human visual embodiment—the images are visual gestalts easily recognized by earthbound humans in their trained, but bodily-perceptual skills.

- Then, if the phenomena are emissions of the spectrum, for example outside our color horizons, translation mediation supplies colors we humans *can* see. Scientists call this "false color" but I call it "relative color" since the implication of a "true color" is frankly irrelevant. What is relevant is the question being raised and its answer in terms of assigned color entails the purpose of the question. Such color assignments function similar to *map* making with colors both falling into a convention, and into an assigned purpose. In astronomy such active manipulation now yields better and more rigorous results than in previous analog mediation. In astronomy a current research frontier is the search for exoplanets [extra solar system planets]. While some 300 have been identified, most are 'inferred' by imaged wobbles and eclipse-like passages across the face of stars. However, a more complex process has recently yielded the first direct images of exoplanets orbiting Formalhaut and Star HR 8799. The process of imaging is a complex one, involving both Hubble orbiting and Keck land based telescopes, and using infrared to mask otherwise obscuring effects of background light and dust and variations upon other enhance and contrast techniques. Thus different 'slices' of the spectrum, including infrared invisible to humans, are used to produce the phenomena of planets in orbit. (7) An older and simpler favorite of mine has been the X-ray image of the pulsar at the center of the Crab Nebula. Here the spinning pulsar with its opposing radiation jets appears clearly in this again bodily unseeable slice of the EMS, but translated into a visual gestalt for our vision. But this structure only shows up in this X-ray slice into the EMS.

- In the cases mentioned I would also claim that the imaging process is one made possible only by the ability to *manipulate* and *construct* the production of the image. Using infrared to mask other effects, computer tomography to vary figure and field, and the use of multiple instrumental variations *cannot but suggest to a postphenomenologist* that these practices are, in effect, *instrumental phenomenological variations*. These are clearly not 'passive,' but 'active' or *genetically* generated instrumentally mediated images. Peter Galison has noted that scientific notions of objectivity change over time. The older, more passive style of imaging— based primarily upon film photography took objectivity to be something like 'photo-realism.' **(8)** This now computer tomographic process is much more akin to material phenomenological variation (and to art making practices as well).

- The parallel between phenomenological variations goes deeper here as well. Each instrumental variation may be taken as a specific 'perspectival' view upon the object imaged. And in this case, when such perspectives reach beyond direct perceptual capacities, one can note that ever larger, smaller, more distant and more extra-horizonal phenomena can be displayed. The "world" is changed. And that is certainly the case in contemporary astronomy. As late as 1920, the dominant opinion of astronomers was that there is one galaxy, the Milky Way, and only after a new measuring system came into practice, did other galaxies become recognized (Andromeda the first). Black holes, pulsars, and myriad objects unknown even in the $19^{th}$ century populate today's heavens. Today, of course, there are millions and billions of galaxies—and to top it off, the very recent remeasurement of our own Milky Way as just expanded by 50%. **(9)**

- One can also add a new feature to variations as well. In phenomenological investigations variations are employed to discern 'invariants' or structures. In the material instrumental variations I am describing, by using such variations one can also see a *convergent* set of results which can, when convergence occurs, be very *robust*. Dating measurements are perhaps the most obvious example with respect to such multiple perspectival instrumentation. The idea is simple: if multiple means of dating measurement can be employed, then if they converge on a narrow date-range, the result may be considered more robust than if only one means is available. Thus if radiocarbon dating, molecular spectroscopy and luminescence dating converge, one can have a much higher

## 5. Beyond Embodiment and a Return    63

probability that the result is robust. In practice, since not all means have convergent calibration ranges, one must often search around for other corroborations. In the search for the history of global warming and cooling trends, earth scientists have often developed ingenious means of extending such histories, such as the measurement of gas ions trapped in polar ice sheets, ocean sediment analysis and many other surrogate sources.

I have now outlined a few of the most salient features of the imaging technologies which produce a second scientific revolution, a revolution which I will now characterize as—in addition to an even more radical extended embodiment, but also as a *critical hermeneutic*.

One of the differences between the first magnificational, compared to the second translational mediations is a difference in how human horizons are recognized. In the first revolution, as in the sound beyond sound example, the embodied-perceiving human discovers the sound beyond sound *within* a wider whole body experience. The feeling of the gong vibrations is immediate and recognizable in a way which makes the analog apparent. This same style of 'aha awareness' also happened in the discovery of the infrared as an extension within the color spectrum. William Herschel, experimenting with prism mediated light, *felt heat* just below the last of the visible red of the visible projected light, thus becoming aware that 'light beyond light' was being radiated. I would call this a recognition of horizontal limits *within* wider whole body experience.

In the case of our 'posthuman' instrumentation of the second revolution, the recognition is more closely tied to perceptions *presented by the instrument*. In a non-astronomical example, Roentgen experimenting with a cathode ray tube, detected a green glow on the screen and when he placed his hand into the beam, discovered X-ray imaging which passed through his flesh such that the inner-bone structure of his hand appeared in the image. In this case what I would term a *critical hermeneutic* interpretation was needed to understand the phenomenon. To 'read' the image in this case is both to *perceive* the image gestalt, and to 'decode' or interpret it in its beyond- experience context, its *translational mediation*. We could now extend this notion to a number of the above examples: the same process applies to the X-ray images of the Crab Nebula or the exoplanet images, etc.

The trajectory I have been following, however, has focused upon

64     5. Beyond Embodiment and a Return

perceptual gestalts in the images. And while this makes it somewhat easier to detect the embodiment transformations, a different kind of imaging shows more clearly the hermeneutic dimension of instrumental mediation. I call the previous 'picture-like' or perceptual gestalt imaging *isomorphic* in that there is retained some obvious semblance between referent and image. A much more obviously hermeneutic style of imaging is what I call *non-isomorphic* imaging. The two most familiar examples of this imaging can be found in the myriad types of *spectroscopy* and the imaging of DNA-RNA chromosome strings. In both cases the images appear as 'bar code' images. In astronomy, for example, such images were produced in the late $19^{th}$ century through telescopes equipt with both some grating and a photographic camera. Only after several decades of experimentation did the observers learn to recognize that the pattern of light (emission) and dark (absorption) lines were actually chemical signatures being emitted from sun, star or other sources. Similarly, bar codes of chromosomes yield the distinctive patterns of GCAT codes familiar to genetic biologists. Note that these 'bar codes' *do not* resemble their referents, but do demand a highly specialized 'reading' or critically hermeneutic skill to understand. For the expert, of course, the patterns displayed are parallel to perceptual gestalts in that the reader recognizes 'at a glance' the meanings of the bar codes. Both isomorphic and non-isomorphic patterns relate to bodily-perceptual skills which are acquired through training and the development of expertise as well described in much of Hubert Dreyfus's work. (10)

**Postphenomenological Ontology:**
Throughout the preceding analysis there lies, just below the surface, two contrary tensions which I have learned only slowly to recognize. I shall call the first 'Galilean.' Insofar as Galileo is often claimed to be both a paradigmatic and tradition setting figure for early modern science. Less recognized as important to this role is the *rhetorical* style which he also established. I will resort to somewhat of a caricature of this style: first, Galileo was primarily 'outward oriented' in his rhetoric. He made much of what he had observed 'out there,' sunspots, the mountains of the Moon, Venus's phases, the satellites of Jupiter. Contrarily, he made little of what I am calling here embodiment or the bodily-perceptual experience within which the observations took shape. He was aware, particularly since observations through a telescope were often doubted and called into question, that one had to *learn* to see through a

telescope and explicitly claimed that "anyone can see what I see" but under the conditions that he first teach the new user how to use the telescope. This is at least an indirect reference to the role of embodiment. The other claim, noted above, was that such telescopic vision was *superior* to unaided human vision (although many of his claims about the differences are now known to be in error, cf. Brown .(11). Both these habits remain deeply engrained in much contemporary science rhetoric—and they are echoed in much of my above account as well. I am an avid and long time reader of both Science and Nature, plus a few other science magazines and I frequently find myself, even though aware of the rhetorical bias, echoing that tendency. Here, for example, in the many examples cited I point to what is discovered 'out there' and in degree echoing the high hopes for the latest and newest instrumentation such as the developing telescopes reported. This, in spite of my own sensitivity and awareness that *virtually no mention* of embodiment or human bodily-perceptual roles occurs in such literature! (And when it is, it is usually mentioned in the negative terms of human limits!)

The other vector which I have emphasized, *is the role of embodiment*. I have claimed that it is a necessary component of scientific practice, plays a role in each of the three types of embodiment noted, and remains an *anthropological constant* in science praxis. Clearly, this vector comes from phenomenology which retains its experiential, and I would say bodily-perceptual emphasis. But, I would also argue, that unlike classical phenomenology, postphenomenology attempts to de-subjectify and to embody as praxis the earlier forms of phenomenology. The notion of embodiment effectively replaces that of consciousness here. This can best be seen by taking an 'ontological turn' towards an inter-relational ontology. **(12)**

An inter-relational ontology has been central in different ways from the beginnings of the phenomenological tradition. In its Husserlian form, however, it remained caught in 'philosophy of consciousness' language and echoed the subject/object problems of early modern epistemology (Descartes) as in the *ego-cogito-cogitatum* language of the *Cartesian Meditations*. The post-Husserlians, Martin Heidegger and Maurice Merleau-Ponty retained the inter-relationality in *In-der-Welt-Sein* and *Etre-au-Monde* formulations. In Heidegger's case the shift was to Dasein, being-here, and in Ponty's case *corps vecu*, both clearly more 'bodily' than

## 5. Beyond Embodiment and a Return

'consciousness' *per se*. Throughout this tradition clearly what had been a metaphysical distinction between subject and object, mental and material, is overcome to the extent that there could be no subject without a world. And while I have come to prefer the parallel inter-relational ontology of Dewey, modeled upon more of an organism-environment, even evolutionary notion, what remains clear in all such ontologies is that there is an implied inter-constitution as the dynamics of the ontology. Put simply, inter-relationality implies that human > < world changes are such that for every change in a 'world' there is a correspondent change in the 'human.' What I have added to this tradition is a sensitivity to materiality and its inclusion into the notion of intentionality itself (as demonstrated in my phenomenology of technics).

Once more I return to Nature's year of astronomy and Owen Gingerich's observations which in this case pair changes in astronomical cosmology and human self-understanding. The usual 'enlightenment' shibboleths depict such changes as a progression of 'displacements': Copernicus, confirmed by Galileo and Kepler, *displaces* humanity's privileged position on earth as the center of the universe; Darwin displaces the deep and privileged position of humans vis-à-vis animals; and later, Freud displaces Cartesian transparent rationality from its god-like role. Such shibboleths are, of course, too simplistic. But they do hint at the easy way to detect world changes in relation to human changes in self-interpretation. We are a long way from the dualistic physics of Aristotle with its distinction between terrestrial and celestial realms, and early modernity slowly won a unification of a 'nature' which operates in the same ways on earth as in distant galaxies. This move has extinguished a hierarchical world and thus places us, not in a privileged, but within the on the same 'nature,' We have also now long lost any notion of a 'center' of the universe. And today, with the increasingly sophisticated means of searching for exoplanets, Gingerich claims that, "The International Year of Astronomy might well launch the next intellectual revolution in our understanding of our place in the universe. Could this have as much of an impact on society as Galileo's and Kpepler's entrenchment of the heliocentric view?" (**13**) This revolution, Gingerich claims, revolves around the question of whether we are *alone* in the universe.

What I want to point out is that for every change in cosmology, there is a correspondent change for human self-understanding. But that is the easy case. The harder case relates to *embodiment*. Insofar as we remain physiologically modern humans, *homo sapiens*

## 5. Beyond Embodiment and a Return   67

*sapiens*, there is a clear, traceable continuity of our physicality with our ancestors of 100,000 < 200,000 years ago. And although there has emerged evidence recently that our evolution continues, and has even sped up in the last 10,000 years, we are 'bodily' still recognizably related to our ancestors. Embodiment, however, as I conceive it, is not merely physiological—although it is also that. It is as Donna Haraway would put it, it is 'natureculture' or 'culturenature' and includes at the least the differing perceptual-bodily skills we can learn. The drastic late modern change in human health with its impact upon life expectancy is but one variable. Who today is really 'old'? Life expectancy in any advanced, industrial country now reaches into the 80's whereas even a century ago it was closer to half that in vast areas of the earth. This phenomenon, too, is deeply inclusive of the technological. I, myself, just turned 75, precisely the age when my father entered the nursing home for his last two years of life. And, I too, might have had a similar fate were it not for state-of-the-art heart surgery. But through several years of medical imaging technology—I now have a pile of CDs with MRI, Echocardiogram, TEE and CT scans which imaged in both black and white and glowing color contrasts image the precise problems of a regurgitating mitral valve and arterial blockage. Thus at the pre-surgical consultation with a multi-screen display, the surgeon was able to point out the problems and describe what he planned to do by way of surgical intervention. So, I am today differently 'embodied' in both a physical and restored motility sense. I did chose the less cyborg alternative of repair over replacement and thus remain 'human,' able to continue with doing philosophy with renewed energy. **(14)**

If my body now technologically restored, and in that sense with a radically different outcome than was possible for early modern humans, so is my world radically different. The alternations of electronic communications with face-to-face communities, the quickly spanned global travel possibilities, in addition to the knowledge base mediated by technologies as well, allows what is happening here and now to be possible and for us to be part of this contemporary world.

**Endnotes:**

1. Owen Gingerich, "Mankind's place in the Universe," <u>Nature</u>, Vol 457, No 7225, 1 January 2009, p28.

2. Don Ihde, "Art Precedes Science, or, did the *Camera Ob-*

*scura* 'invent' Modern Science?" in Petran Kockelkoren, *Mediated Vision* (Art EZ Press, 2007), pp. 24-37.

3. Gingerich, Nature, p. 28.

4. Harold I. Brown, "Galileo on the Telescope and the Eye," Journal of the History of Ideas, 1985, pp. 487-501.

5. See my "Husserl's Galileo needed a Telescope" PDF on the Stony Brook Philosophy home page.

6. Don Ihde

7. Science, Vol. 322, Issue 5905.

8. Peter Galison, "Judgement against Objectivity," in Galison and Jones, *Picturing Science Producing Art* (Routledge, 1998), pp.327-359.

9. Science, *op.cite*.

10. Hubert Dreyfus, *What Computers Still Can't Do* (MIT Press, 1993).

11. Brown, *op cite*.

12. See Peter Paul Verbeek, *What Things Do* (Pennsylvania State University Press, 200 ) and Evan Selinger *Postphenomenology: Critical Companion for Ihde* (SUNY Press, 2006).

13. Nature, *op. cite*, p. 29.

14. Don Ihde, "Aging: I don't want to be a Cyborg," in *Ironic Technics u*(Automatic Press, 2008), pp.31-42,

# 6

# IT: Clouds and Cyberspace-Time

**Predictions:**
The Economist, 25 October 2008, claims, "Computing is fast becoming a 'cloud'—a collection of disembodied services accessible from anywhere and detached from the underlying hardware... everyday computing will one day be mediated by this ethereal cloud..."[1] The article then goes on to characterize this 'cloud' in contrast to its 'locations:'

- "...computing will be a borderless utility... it need not matter whether your data and programs are stored down the road or on the other side of the world...

    But everything will look as if it is happening on the screen in front of you...

    Yet geography still matters. The data centers that contain the cloud... cannot be built just anywhere. They need cheap power, fibre-optic cables, a chilly climate and dry air... Iceland fits the bill. Google is even thinking of building floating data centers, cooled by seawater and powered by waves...

- The legal and political issues... more than previous cross-border utilities... will be a cosmopolitan prisoner to laws that are mainly local."[2]

Admittedly, one should be skeptical of any predictions built upon even state-of-the-art technologies. It is not at all sure that fibre-optics, the need for cooling, and favored geographies will remain the constraints on future IT, but these predictions do outline some of the anomalies we currently face in the experience and use of now globalized computing.

My strategy here will be to begin with practices which should be familiar to all, or at least most, of us, and undertake what I shall call a *postphenomenological analysis* of the history and phenomenology of a set of radical changes which have taken place primarily since the late $19^{th}$ century to the $21^{st}$. I shall employ variational theory to these changes, with focuses upon both our embodied experiences and upon the material or technological embedding which transforms our practices. In short, this will be a postphenomenological topography relating to our new computational or IT developments. I begin, however, with our shared common practice of *writing* which I am assuming all academics and legal persons have familiarity.

**Variational Moments:**
Michel Foucault's "What is an Author" makes a series of observations about how what he calls *the author function* changes with different historical discourse *epistemes*. For example, one author-function, now in a surpassed *episteme* would attribute some text to an authority figure which then would give it weight. But a text attributed to some classical figure—Aristotle, for example—or even more likely a text attributed to some sacred or religious figure—Moses for the Pentateuch, or St. Thomas for philosophical treatises—clearly enhances the importance of that text. By later modern standards such attributions are, at the least, questionable. No critically informed Biblical scholar today would hold that some individual Moses, wrote the first five books of the biblical canon. And, while I find it ironic that a vestigial 'authority aura' remains a ghostly presence in so much generic continental philosophy, for the most part this author-function is relegated to a lost past.

Foucault claims that the modern sense of the author-function whereby a text is assigned as belonging to some actual individual author when, "[Texts] are objects of appropriation, the form of property they have become is of a particular type whose legal codification was accomplished some years ago... Speeches and books were assigned real authors, other than mythical or important religious figures, only when the author became subject to punishment and to the extent that his discourse was considered transgressive... it was at the moment when a system of ownership and strict copyright rules were established (toward the end of the eighteenth and beginning of the nineteenth century) that the transgressive properties always intrinsic to the act of writing became the forceful imperative of literature."[3]

So typically Foucault! This author-function, individual author owning his/her text, is thus a modern *invention*. And, if invented, then its possibility for dis-invention also is part of the Foucaultean passing of *epistemes*. Note for the moment, that the Foucaultean description is focused upon discourse and social-political practices and does not reference technologies of production, mediation, or transformation.

A second variational moment, one from Martin Heidegger, does situate what Foucault has called an author-function, with *writing technologies*. In his *Parmenides* Heidegger claims:

Human beings 'act' through the hand: for the hand is, like the word, a distinguishing characteristic of humans. Only a being, such as the humans that 'has' the word (*mythos,* logos) can and must 'have hands.'... The hand has only emerged from and with the word. The human being does not 'have' hands, but the hand contains the essence of the human being because the word, as the essential region of the hand, is the essential ground of being human."[4]

Now, while this passage does not yet mention the *pen,* it is clear that the Heideggerian practice is composition by pen. A marvelous little book, *Heidegger's Hut* (2006), shows Heidegger at both his Freibourg house and in his hut, with pens, blotters and ink bottles, a manuscript laid out on his desks. And the description above is almost a paean to his own practice. The clincher, however, soon follows, again with Heidegger's typical disdain for the modern:

It is not by chance that modern man writes 'with' the typewriter and 'dictates'—the same word as 'to invent creatively into' the machine. This 'history' of the kinds of writing is at the same time one of the major reasons for the increasing destruction of the word. The word no longer passes through the hand as it writes and acts authentically but through the mechanized pressure of the hand. The typewriter snatches script from the essential realm of the hand—and this means the hand is removed from the essential realm of the word.[5]

It is clear that for Heidegger, 'authentic' writing is *penmanship*. He privileges what was certainly a long history of writing technologies, from quill to fountain pens which began to be displaced only in the late $19^{th}$ century, first by the typewriter, then on to word processing in the $20^{th}$ century.

My third variational moment also occurs at the historical juncture when electronically mediated 'writing' begins to exist at least alongside, but sometimes displacing both penmanship and typewritten practices:

When the project to computerize the commentary on Jewish law got underway at Bar Ilan University in Israel, the programmers faced as puzzle. Jewish law prohibits the name of God once written from being erased or the paper upon which it is written from being destroyed. Could the name of God be erased from the video screen, the disks, the tape? The rabbis pondered the programmers' question and finally ruled that these media were not considered writing: they could be erased.[6]

Here is an interesting compromise—the rabbis could in some sense be said to remain with Heidegger and claim that 'authentic' writing is what takes place when a scribe *inscribing with a pen* makes an inscription upon paper, vellum, papyrus. But simultaneously, they did not outright *reject* the new media which has, one might say, an essential feature: erasability.

**A Short History of Writing:**
The three variations just noted are but instants in a much longer history of writing. Writing itself transformed cultures, from oral to literate, with interesting shifts I will not explore here. And, we are also currently in an epoch in which the older master narrative regarding origins of civilization is being challenged and modified. That narrative, which originates the 'oldest' of many human cultural phenomena in Egypt, Mesopotamia on into Greece and Indo-Europe now has to give way to equally old or even older origin stories across a more distributed ancient global world.

If writing begins with meaningful inscriptions of multiple kinds: pictographic, wedge-shaped as in cuniform, abstract hatchings, then writing or proto-writing is very ancient indeed:

- Where and when humans 'first' began meaningful inscription is an open question. I, myself, have viewed some of the graphic icons in Ice Age caves, and have seen calendar and lunar markings on reindeer antlers going back longer than 20,000 BP.

- An explosion of inscriptions emerged in Egypt, Iraq and Pakistan between 5600-5200 BP;

- Romania, with an older Sumerian culture, now claims inscriptions back to 6500 BP;

- And with China, a previous turtle shell inscription process first thought to go back roughly 3500 BP, now has examples dated to 7000 BP

- And even in the New World, Olmec inscriptions, can be dated back to 3000 BP.

I will not take sides here, but will claim that our 'histories of writing' have necessarily become more multicultural and ge-

## 6. IT: Clouds and Cyberspace-Time 73

ographically widespread. But what intrigues me more are the *apparent similarities in style and materiality* in these early practices. First, the final material 'tablet' as I shall call it, is something **hard,** upon which is **etched, inscribed** some set of simple **figures.** In the cases of Mesopotamian cuniform, the tablet is first clay onto which are inscribed wedge-like figures; in the case of the Pakistani Harappa inscriptions, it is also pottery-like, both baked and made into the hard product. Others, etching onto stone (Olmec) or turtle shells (Chinese) have a similar productive style. All this is fortunate for us, of course, since such a practice produces long term durable results. Secondly, the stylus or other instrument making the inscriptions was also **hard,** needed to make an actual inscription.

### A Phenomenology of 'Stylus Writing':

Grouping now the styles of writing which are similar: an embodied human—using some sort of inscription device, a 'stylus'—produces an inscription upon a 'tablet' whether originally soft, but once processed, hard. Those of you familiar with my earlier analysis of a phenomenology of technics, will recognize here an *embodiment relation:*

Human-technology (stylus)-tablet

The human using the stylus projects through it the project and makes an inscription. Here is bodily action, in this case somewhat similar either to pottery making (if the tablet is soft) or sculpting (if the tablet is hard). Effort is involved; there needs to be the trained skill developed to the constraints and capacities offered by the stylus; training is needed to attain the 'transparency' of embodied tool to produce the recognizable marking which constitutes the 'writing.' The stylus is, of course, a technological instrument *through which* the action of writing occurs. But, this instrument is non-neutral—some actions performed through it are easier than others. For example, making straighter marks with either a cuniform stylus—or much later a Roman chisel making phonetic inscriptions upon stone—allows straight cuts more easily than curved ones, such that I, V, W, are easier than O, U, S. Then, too, if one has a long and complex text to produce, speed is called for and one can in many, many written languages detect how abbreviations, an evolution toward a simplification of script, takes place. The stylus does not *determine* this, but the relative ease or difficulty *inclines* the user to often take paths of least resistance. All this is minimal, but hopefully suggestive. I am suggesting that these early forms of writing bare parallel experiential

74    6. IT: Clouds and Cyberspace-Time

features similar to pottery or sculpting practices which at the least are slower and more difficult to accomplish than later practices. And although I doubt anyone would be inclined to write a cuniform tablet letter to a friend today, were one to try, one could re-encounter the relative difficulty of such composition. And, think too of the temporally long and complex process of inscribing a clay tablet; baking it in a kiln; and finally sending it by some slow messenger to the recipient.


**Pen and Ink, Softer Writing:**
I will now make a multi-millenia leap to an epoch of what I shall call 'soft tablets' or variants upon pen, ink and paper. Here the stylus, previously a hard, often pointed instrument, gives way to brushes or early pen-variants which while retaining hard handles, now come with 'softer' tips such as brushes or with pen-like tips as in slotted quills. The previous pottery, stone or ceramic tablets now give way to paper-like sheets made of parchment, papyrus or paper. Chinese calligraphy which goes back to 4300 BP, or Egyptian papyrus writing 3000 BP, as examples. Again, concentrating upon materiality, both stylus and tablet take 'soft' or flexible trajectories of human bodily actions. We retain the embodied human, mediating writing action or practice through the brush or pen. As previously, the "hand" still uses the stylus, now become flexible-tipped. New bodily skills are needed to produce the flowing styles of such instruments. The result, perhaps no longer an 'inscription,' but inked or painted figures are now produced upon the now 'softer' tablet. The same features concerning the learning of skills remain, the accomplished calligrapher or skilled pensperson needs to have 'embodied' the capacities and constraints of the material medium. And while a critical phenomenology of this changed practice would reveal many changes from the earlier pottery-sculpting mode, just one change should be suggestive: Brush and paper and flexible pen and paper *reverse* the previous ease and difficulty pattern for the resultant inscriptions of the previous technologies. Here, curves and circles are *easier* than strict and straight lines! Just look at calligraphy and some example of *belles letters* and you will see what I mean. These variations upon pen, ink and paper, like their forebears of stylus and tablets of clay and stone, also carry a very long history—indeed up to Heidegger in the $20^{th}$ century and including the period of Foucault's literary text owning author.

## Keyboards:

Another millennial plus leap, enter, at the very end of the $19^{th}$ century, the keyboard, first as typewriter. As Friedrich Kittler has pointed out, Nietzsche loved the keyboard and saw it as a distinctly improved writing technology—especially for poetry.[7] One of the unpredicted social side-effects of the typewriter was a massive gender change within the cadres of secretaries. Many males, who had dominated the role of secretary, saw the typewriter as a degradation of their acquired skills and stepped out of the profession. But younger women, already pre-skilled by way of bourgeois piano training for proper young ladies, filled the gaps and soon dominated the secretariat.[8]

Phenomenologically, however, in this new human-writing technology-text intentional arc, the typewriter replaces the ancient stylus *and* the pre-and early modern pen or brush. The tablet, having become paper, remains paper now rolled through the mechanical carriage of the typing machine. The mediation of writing is now produced *through the keyboard* rather than the stylus or pen. Here, however, Heidegger's description fails a phenomenological descriptive test—he obviously was never a skilled typist. His claim that the word no longer passes through the hand, but becomes mechanized as pressure on the hand with the script 'snatched' from the hand, reads to me like a beginner's description where one has not yet attained the flow from the hand transparently through the machine onto the paper. A high speed, skilled typist feels the flow from playing the keyboard to the typed script on the paper. Of course the very hold is different, rather than grasping a single stick-like stylus, later pen, grasped by the hand, one now uses the whole hand with all fingers brought into action. One still produces writing via the hand, although now the hand spread out over the keyboard. This flow is also phenomenologically similar to the skilled pianist who, from hand through the piano, produces a virtuoso piece of music. As for literature and typewriting, from Mark Twain on through Hemingway, most $20^{th}$ century writers had switched to the typewriter for original composition. And as in the Heidegger moment, many writers praised the quickness and ease which the typewriter allowed. For example, one follower of Hemingway claimed, "At some point my obsession with Ernest Hemingway led to a love and appreciation with typewriters. Nothing is more pure than a typed letter on a manual typewriter."[9] From the manual typewriter there followed a very quick evolution of the keyboard as a late modern writing technology, which moved from the mechanical typewriter to the electric typewriter to the electronic keyboard, and with each transformation one follows the technological trajectory of making the task easier and quicker to undertake. In my own case most of my early books were composed on a mechanical typewriter, then an electric one, and by the mid-eighties a word processor. Today I find it difficult to even use a mechanical typewriter! I am suggesting that developmental trajectories may be hard to reverse. And while I know of a very few colleagues who continue to first 'pen' their manuscripts before word processing them, I know of none who today who still use a mechanical typewriter.

**Excursus on Dissemination:**
In my narrative I have focused upon what I take to be our common experience in writing composition rather than upon the equally important history of dissemination. After all, once baked, printed, or published, the written works could be copied, circulated, disseminated. There remain thousands of cuniform tables in Mesopotamia, now in museums; libraries have collected thousands of manuscripts and bound books. And auctions of hand-written letters of famous people sell for high prices.

I shall not closely follow this process, but it is easy to note that during the era of baked, hard tablets, camel and donkey trails through mountains and deserts, or with runners, had to be a slow and often dangerous engagement. By the time of Foucault's modern author-function in the eighteenth century on, railways and sailing ships could be used to transport written works along with the postal systems not too very much earlier. This history is also part of the transformation of space-time which precedes cyberspace-time.

**The Visual Display Screen:**
My own first word processing technology was a 1984 IBM PC, which when I junked it some years ago my son protested that I had trashed "a museum piece." With it the practice of writing again changes. In the immediate use, the *screen* now becomes the tablet, first in glowing green letters, then amber, now the full color range. In the composition process the screen substitutes for paper. And, as already noted by the Rabbis, words can appear or be erased with equal ease. Phenomenologically the transformation of the tablet-paper function to a screen in writing praxis is quite radical. But allow me here to take stock of the story so far:

## 6. IT: Clouds and Cyberspace-Time 77

- Throughout this "history of writing" the intentional arc of human-writing technology-text may be discerned. At each stage, the human is bodily engaged in producing the written result. The author, writer, first must learn the skills which makes the written text possible—and the skills are somewhat different in the literal inscription styles, mimicking in some sense pottery or sculpting techniques. The shift, with the turn to brush and pen variants, with inked patterns placed upon paper, parchment or papyrus, entails different bodily actions, but still these have to be learned. Then from a hand holding either a stylus, brush or pen, with the keyboard, whole hand movement must now be learned.

- The 'mediating' technology also lies within the intentional arc, and when through skilled and learned bodily action it becomes 'embodied' quasi-transparently, can allow the 'flow' to occur. Each variant, however, poses a different set of capacities and constraints, or as Andrew Pickering calls them, accommodations and resistances.[10]

- The result, then, is the appearing text, first as inscribed and made 'permanent' upon the tablet or pottery, then with less permanence, but still a long temporal continuity on paper, parchment or papyrus, and finally, the equally easily produced but also easily erased screen texts of word processing.

I am claiming that throughout this vast history of writing, what I call embodiment relations—*through a wide variety of cultural-material variations*—obtains. But, now having finally arrived at out own era, I will begin a deeper phenomenology of, first, the screen and its multistability.

### Multistability in Cyberspace:
Once the technological mediation of writing takes place on a screen rather than on a tablet or paper, we reach a horizon for written words. This form of composition is clearly more *flexible* than the previous processes. While it is true that placing letters on a screen by bodily hand motility, is both easily accomplished if a skilled typist, and equally easily dissolved or erased through deletion, cutting, and other keyboard actions, one should first note that yet other key punches, *save* or *print* produce a more permanent result. In other words what we today call *hard copy* is one additional step from the older typewriter, pre-screen technology. In this delay or extra step in producing a *hard copy* result, much

## 6. IT: Clouds and Cyberspace-Time

romantic nonsense has been also inserted—only the hand produced text (Heidegger) or the permanent result (the Rabbis) is 'authentic' writing. This seems silly to me since the 'hardness' of the cuniform tablet also called for baking before it became permanent. Instead, what word processing does, is to relocate where and at what stage reflective editing can occur. In slow handwriting I can make decisions about changes along the slow way; in old fashioned 'cut and paste'—once a more literal process with pre-correction technology typewriters—the process was differently located. But with screens the revisions and reflective editing can occur at many points, and with ease. In other words, computer based keyboard plus screen is in practice *flexible,* even more so than pen or typewriter and paper. But, it is also *multistable.* The introduction of the screen produces a technological *hybridization* which transgresses the boundaries of previous writing praxes.

- First, staying with writing with word processing, the screen can be said to have an 'opaque' stability. As I write, the words take shape on the screen before me—or, if I am reading a text, the words are already arranged as if printed out. The 'opacity,' however, is a lighted opacity and in a variation upon ancient scrolls the page 'rolls out' as it were. I shall call this the *reading stability.*

- Screens, however, can also be used to watch movies, videos, or be *interactive* as in playing computer games. By placing the screen into something like a tablet-paper position, one brings to this new position the hetereogeneous capacities of screen histories which include cinema, television, DVDs and the like. Here the flexibility of the visual display screen can yield to a 'transparent,' even 3-D appearance. What one perceives is not on the surface, but is seen *through* the screen, beyond the screen—now commonly called cyberspace. In interactive mode, the player—who can use the keyboard, but who can also substitute yet another variant, the joystick—can 'enter' this space as a player. I shall call this *playspace stability.*

- Playspace becomes even more multistable since there are choices of 'points of view' or POVs. I can assign myself an 'embodied' position so that if I am carrying some weapon, it appears in front of my presumed or virtual position and I can then follow the various pathways which appear in the cyberspace through the screen, or I can assign myself a 'disembodied,' or *out there being,* in which I identify with some figure perhaps riding some kind of science fiction vehicle, and again now 'see myself' shooting at the bad guys. Or, in a sort of hybrid, I can have this out there being close to me in my seated position as if I am 'piggy backing' my alter-self. In short, this style of screen flexible multistability allows me to chose different bodily positionings. Phenomenologically, of course, it should be pointed out that there are basically two perspectives involved with constituting game POVs: With the embodied position, with the gun extending immediately in front of me, my actual visual position is privileged, but then note that the game 'world' is what moves in cyberspace, whereas in my seated position I remain stationary. Switching to the disembodied or out there position, my perspective becomes a 'bird's –or god's eye- perspective whereby I am viewing the scene from 'outside or above.' Recognizing the artificiality of such sitting positionality, the latest more bodily active Wii games now call for limited bodily action as in returning tennis balls or bowling virtual games.

Again, I am suggesting how screens change the older writing paradigms in significant ways due to the flexible and multistable displays screens can present. And with this flexible multistability, all of which takes place at the computer station, *there can be no privilege to writing per se!* Word processing may well now be the dominant way writing occurs, but it now occurs in a different and transformed context. To this point, however, I have been focusing upon the varieties of human-technology-world practices in abstraction from a much wider, and now global space-time context. The computer station is not simply that of a typing station—it is a locus where human and world interact globally.

**Enter the Internet:**
I have already referred to my 1984 desktop, then not yet hooked up to the internet. The next year I became a Dean and my university was just in the process of computerizing all administrative offices. Heretofore, the SUNY system had not allowed purchase orders for computers, but once making the move, in a few years they reversed priorities and no longer allowed requisitions for typewriters. In the transition a goodly number of older secretaries retired, finding the transition to the new complicated "typewriters" with their clumsy word programs simply too much to cope with. For a

while it appeared that a reverse 'Kittler' move might occur with female secretaries leaving and younger male computer jocks replacing them. Now, however, no matter where one goes, walking by faculty offices, into departmental offices, past vast rooms for student use, virtually everyone is sitting at a keyboard in front of a screen.

This same era brought the internet, touted to bring on the "paperless society," but at first it was a struggle to determine if the internet was more like a telephone, or a letter or memo producing process. Email tag replaced phone tag while memoranda as attachments succumbed to the ease of erasure by deletion. And there may be nuanced cultural differences: Americans seem to follow more the phone model, with short messages quickly responded to, while Europeans—in my experience—are more likely to do letter-like responses with longer lag times—although that also seems to be changing. Both are now into short, very informal exchanges, the stylistic center of gravity for e-communication. This was a beginning, but it already began to effect social changes:

- One effect was to download previous tasks undertaken by service personnel onto their own supervisors. Secretaries no longer produced letters, Deans had to write their own. Later we found ourselves doing our own travel arrangements, book ordering, and a plethora of tasks previously done by assistants.

- Another effect was to quickly find "the world itself at one's computer station." Positively, international e-correspondence, from Chinese graduate students asking questions, to peers from other countries organizing conferences (such as this one); negatively, I get spam and scams in Russian, French, German , Hebrew and many other languages.

- Research resources are now available online and today's undergraduate spends far more time online than in a library. Even doing this paper, I take time out to check datelines for events to be noted. So much science publishing in particular now is first in *preprint*. This is virtual, but *real* intellectual globalization. Similarly, actual travel, conferences and research events are arranged first by email, then fulfilled in the flesh. In my own case, by far the dominant first introductions are electronic, only later face-to-face.

With the internet one could say the "screen speaks out to us" just as we to the screen. It is the interactive interface where the cyberworld is mediated. We, in other words, are back to the opening predictions from The Economist. We remain situated and embodied human beings and our locus of this experience is at our *station*. Invisible to us are the satellites, the micro-wave facilities and the routing networks, the infrastructure which supports this now global interconnectivity. And while I have used examples familiar to academicians, the global linkage in every domain is obvious. Most of us publish through overly expensive European academic publishers which give us the options of open access e-publication (for a $3000 fee), color images (for a $900 fee) as well as a later standard **hard copy** paper publication,( although many in journal form also appear first electronically for free to subscribers). My first book on the philosophy of technology. *Technics and Praxis*, as early as 1979 went into copy editing production in Holland, then off to India for compositing, back to me for checking and a second round-the-world trip before final publication. Today the same process is accomplished electronically and my recent book, *Ironic Technics* (2008),is distributed and sold only through electronic sales on a global basis.

The invisibility of this infrastructure as with the predecessor technologies of telephone routing systems is such that the electronic internet shuttles its signals, globally, in decentralized, distributed pathways which include both earth and space nodes, themselves above our very atmosphere. And, as with The Economist, our storage may be anywhere. But, the experiential locus remains the screen at the computer station which we occupy, although sometimes in our office, our home, or while away at hotel or university sub-stations.

I have already noted that in this situation the writing model is exceeded—writing becomes only one of many options at the station. Our stations are far richer than previous writing models, since a full panoply of audio-video phenomena sound and display before us. From the opaque text to 3-D video to cinema and/or music presentations the station performs its multifunctions. To me, in addition to the displacement of previous text-dominated screenings which must be but one variant alongside the full range of audio-visual displays, the connection to the latest scientific imaging moves our station located experience to 'worlds' which beyond our perceptions now can be translated for our perception. I refer here to contemporary imaging technologies.

Yet it is at this locus of experience that the cyberanomolies occur. It is here that cyberspace-time takes its shape. First, spatiality: In contrast to geographic space, cyberspace is always the *nearly same* near-distance. Whether the email is from China, the news clip from Mumbai, it is here at my station, where I am usually seated and using eye-hand motions that I experience the voices and messages from elsewhere. With today's electronic IT, the old noise factors for earlier technologies have largely been overcome. Early telephones, for example, had a signal/noise ratio such that by listening one could tell if the call was from afar partly by virtue of the static or noise heard. And, if from very long, say transatlantic cable transmissions, one could also detect a *time-lag*. Today's cyberspace-time is much more transparent and presents a much more uniform near-distance, although the time lag phenomenon remains as in the pauses between television anchors and correspondents in the field half a world away.

It also remains the case that so far cyberspace-time presentations remain partially *reduced sensory plenum* relations, audio-visual. I can recognize your voice or your face, but I cannot shake your hand, even though you remain the recognizable mediated you. One can switch between multiple modes of mediation—text and text messaging, voice via *skype*, audio-visual in teleconferencing, etc. And when these modes are intercontinental, the vestigial trace of space-time distance remains in the time lags. For the most part, however, space-time is *compressed*.

If one now moves from the relatively rarified realm of IT, space-time mediations of a much more material sort also can be carried out. The November 25, *New York Science Times* pointed out that many of the drones used in both Iraq and Afganistan are controlled from a command post in Nevada! Similarly, 'Nintendo surgery,' as it is called, can be done robotically from equally geographically distant locations. And, even more remote, control of the Mars Explorer and Sojourner Truth, robotic vehicles on Mars, are also cybercontrolled from Earth. In all these cases time-lag remains the experiential trace of geographical space-time. Such time transformations,, often of micro-phenomena, are critically recoverable. Take the example of gaps between one's own time-location and that of your interlocutor. Prior to a recent trip to Japan, my host and I did our communications mediated through a 12 hour time zone difference. Thus we chose times when each of us would be awake with 9:00 convenient, although one of us was time-located in the morning, the other in the evening. This same time/locale difference not located in a time node of awakening time, would then revert to roughly a day's difference. I call this *layered time*. It is flexible and relative and I would even go so far as to say such time layering is experienced as a kind of "Einsteinian time" where space-time displays both its spatially distant, temporally located structure. The *materiality* of temporality is felt. Clearly such a space-time is neither *linear* nor *universally uniform* time. Thus in contrast to earlier eras, time and distance are now more rapidly traversed as a near *hypertime*.

**Alternations:**
Today's world, at least in all developed areas, has increasingly many sites which have stations where situated, embodied humans interface with the mediated IT world. At my university many buildings have *Syncsites* where rooms are filled with computer screens and keyboards at which sit the rainbow of students at all hours. My Chair leaves his door open as a welcome sign, but he sits at the opposite end of his office usually at the keyboard/screen and often with telephone also cradled on his shoulder. The office staff are usually online as well. In the town there are internet cafes, WIFI hotspots, and so one could conclude that all are interconnected. But as a philosopher I am always wary of any slippery slope tendencies. At our stations, unlike the exaggerations of science-fiction, we are not permanently 'wired-in.' Rather we move about in our compoex world through flexible *alternations*. At least a year has now been spent preparing for this conference. A few of you—Mireille and Peter Paul—have been known to me from previous pathways and we have crossed and met face-to-face in addition to multiple e-crossings.

But most of you only now become fully face-to-face, situated and embodied with all of us here together. I am not suggesting at all that we have a sort of 'double' world or worlds, not at all since we are fully situated and embodied when at our stations too. I am suggesting that our lifeworld geography increasingly includes the electronic furniture with which we live and move and have our being. We have begun to notice that insofar as the humanly more dominant activity of sitting at stations carries a price as well, since the lack of exercise and restricted bodily motion, especially when coupled to eating habits may incline our trajectory towards obesity! I have even been a consultant on such a problem with Learning Lab Denmark in a project which involves inventing ways to stimulate and entail bodily activity via computer games (with ideas similar to Wii results). This awareness indirectly responds to the observation above that most IT activity remains restricted to less than whole body activity with keyboard/screen technology below the full sensory plenum. We have yet a long way to go.

**Endnotes:**

1. The Economist, 25 October 2008, p. 17.
2. *ibid.*, p. 17.
3. Michel Foucault, "What is an Author?" in Paul Rabinow, *Foucault Reader* (Pantheon, 1984), pp.101-120,
4. Martin Heidegger, *Parmenides* (Klostermann, 1982), pp. 118-119.
5. *ibid.*,p. 119.
6. Cited in Michael Heim, *Electric Language* (Yale University Press, 1987), p. 192.
7. Friedrich Kittler, "The Mechanized Philosopher," in Laurence Rickels *Looking After Nietzsche* (SUNY Press, 1990).
8. *ibid.*
9. FlickrFind/Writing Machines, www.acontinuouslean. Com.
10. See Andrew Pickering, *The Mangle of Practice* (University of Chicago Press, 1995)